Designing and Engineering Economic Tramways

Designing and Engineering Economic Tramways

Professor L.J.S. Lesley

Published by Emerald Publishing Limited, Floor 5, Northspring, 21-23 Wellington Street, Leeds LS1 4DL.

ICE Publishing is an imprint of Emerald Publishing Limited

Other ICE Publishing titles:
Ground Vibrations from High-Speed Railways: Prediction and mitigation
Victor Kyrlov ISBN 978-0-7277-6379-2

A catalogue record for this book is available from the British Library

ISBN 978-1-83708-399-2

Cover photo: shutterstock.com/g/netfalls
Commissioning Editor: Michael Fenton
Content Development Editor: Ryan Molyneux
Books Production Lead: Emma Sudderick
Project Manager: Rajachitra S

Typeset by KnowledgeWorks Global Limited
Index created by David Gaskell

Printed and bound by CPI Group (UK) Ltd, Croydon, CR0 4YY

Contents

Foreword

The benefits that modern tramways and light rapid transit (LRT) bring to the urban transport scene are well established. Cities that enjoy these benefits can take them for granted. A reminder of such benefits is important, especially when contemplating investing in a new line. Areas that do not currently enjoy these facilities will benefit from investing in an LRT line.

This book is aimed at bringing the many facets of developing an LRT system into one volume. It does so in a thorough and comprehensive manner. The book can be read cover-to-cover (it is quite easy to read) or used as a book of reference on a particular point or issue.

While LRT is a most comprehensive public transport system, offering wide transport, environmental and social benefits, it may not be the best solution. Other less capital intensive modes may provide a quicker solution for a city's transport problems.

When an LRT system is chosen, it is a whole system in one go: track, propulsion, vehicles, control and regulation, fares and ticketing, as a complete package. This book examines each of these aspects and their interaction, as well as their relevance.

Having spent a lifetime in transportation and having planned, engineered and operated all modes of urban transport, I feel well qualified to comment on the relevance of LRT as a means of transport that will bring substantial benefits. This book brings together all the issues surrounding LRT, and will enable people – local and national politicians, professions, user groups and investors – to promote such a system.

A modern tramway or LRT system combines the best of both worlds with the accessibility and high frequency of buses, and the reliability, speed and comfort of rail. These benefits can be enhanced where LRT is segregated from other traffic. There are many cost-effective ways of achieving this, as set out in the book.

Most LRT systems introduced in the UK have been developed by those with an engineering or operating background. Many have privately researched LRT around the world, and most were self-educated. How we wish we had this book, containing years of technical, commercial and practical experience needed to deliver successful systems.

In Britain we regard light rail schemes as 'narrow' transport projects. In many parts of the world, particularly France and North America, LRT is seen as improving the urban realm and helping to regenerate cities with social enhancement. This broader approach helps to 'soften' the raising of the capital cost. In Britain, there tends to be a short-term view about (public) projects. This may help to explain Britain's slow progress in the light rail field.

There are some pre-requisites to enable LRT systems to be developed, through the planning, legal and financial stages. Firstly, there should be an over-arching political structure that will command authority and give the backing and focus to promoting LRT. New larger authorities created in France for agglomerations of major towns and cities are a good example. Secondly, funding structures are important. Some countries – for example, Germany – have a multi-tiered government structure, with federal, land, conurbation and city levels. In the USA, federal and state funds can be augmented by sales and gasoline taxes to fund the investment cost of LRT schemes. In Britain, public funding comes only from two sources: national and local government.

Ideally, a new LRT system needs a 'champion' to encourage and push the project through all its stages to opening and running a service. This can be a councillor or a mayor providing the necessary drive and focus. This is essential to maintain agreement between all parties and the various districts, as well as keeping the public, funders and central government on side.

D Scott Hellewell FCILT,
Former Director of Operation, South Yorkshire Transport Ltd.
Former Operations Director, Manchester Metrolink,
International Transport Consultant (retired)

Preface

A textbook always presents challenges, as it is not read like a novel. Students read sections for a particular piece of information for a project, or to understand the operation of a key equation. Chapters, therefore, have some overlaps with information repeated to avoid the need to refer back (or forward). This should enable a fuller understanding of a particular point. For the reader who treats this like a novel, that repetition may be boring. The textbook is aimed at technical and professional readers, providing a complete picture without having to refer to different sources of information or necessarily reading right through. This will provide an approach to delivering a satisfactory and efficient tram system, at the most economic capital and operating costs.

This textbook is for the English-speaking world for the design, engineering, transportation and system specialists. The author has attempted to use the nomenclature found in different places. He has, therefore, also tried as far as possible to use both imperial and metric units. Several technical terms, with variants, are provided for different countries to gain a better understanding. The principles should be understood by all, even if not all the jargon is immediately recognised.

Between 1968 and 1983 the author's research concentrated on finding ways to attract car commuters to urban bus services. Significant volumes of research worldwide were reviewed. Projects were undertaken while at the University of Newcastle upon Tyne and in Liverpool Polytechnic. The conclusion was that, in countries with high levels of car ownership, there is a small probability of significant modal transfers from private cars to bus services, even to 'smart' or guided 'trambus' systems.

The literature and experience shows that car commuters can be attracted to rail services. The ideal for many cities is a metro/subway/underground network. The cost and disruption required for construction means that installation progress is slow. Building underground is expensive and limited by the availability of capital funding, even when potential schemes can be shown to be profitable.

Tramways/light railways also attract trips from cars. Data available shows that between 25% and 45% of passengers have left a car at home and switched to use light rail or tramways. This point was emphasised by Pierre Laconte, then Secretary General of the International Public Transport Association, who observed 'tramways provide 90% of the benefits of a metro at 10% of the cost'. This textbook therefore concentrates on practical ways to deliver economical tramways, to maximise the role of public transport and provide an acceptable alternative to private car use. This is a tool for urban planners to reduce the environmental impacts of high levels of car usage, toxic air pollution and congestion.

LJS Lesley, 2024

About the author

Professor Lewis Lesley CEng, FRSA, FICE is Technical Director at TRAM Power Ltd. He is an acknowledged expert in the field of urban public transport, has published more than 200 technical papers in refereed international journals and is author or editor of 25 books, including *Light Rail Developers' Handbook*. He has held posts as Public Transport Officer, Durham County Council; Professor of Transport Science, Liverpool John Moores University; and Technical Director, TRAM Power Ltd. He has acted as a consultant on projects in the UK and internationally, and has also been a visiting professor at Budapest Technical University, Delft Technical University, National University of Ireland in Maynooth, Szechenyi Istvan University, Rice University and Leeuwarden Hogeschool.

LJS Lesley
ISBN 978-1-83708-399-2
https://doi.org/10.1108/978-1-83708-398-520261001

Chapter 1
Introduction

This chapter discusses the context and environmental effects of urbanisation and the need for 'mass transit'.

1.1. The urban context

The world is rapidly urbanising, with fewer people living in rural areas (Figure 1.1), as jobs and economic activities are attracted to towns and cities that, consequently, are getting larger. The United Nations estimate that already more people live in urban than rural areas (57% in 2022) (UNCTAD, 2022). In developed countries the urban population is already 80%, whereas in less developed countries the figure is only 36% but growing rapidly.

The nature of urban areas is that people live separately from their place of work, education, shopping and leisure activities. Transport is therefore important for smooth, economic and efficient functioning. Walking is limited by the short distance that most people will undertake, typically up to 1.6 km (1 mile). Some cities have provided cycle routes, and cycling greatly increases the distance that can be covered comfortably, typically up to 8 km (5 miles).

Not all residents can walk or cycle, either due to physical conditions or being parents with small children. The ideal for many people is a personal car. With high levels of car ownership, full

Figure 1.1 World population trends

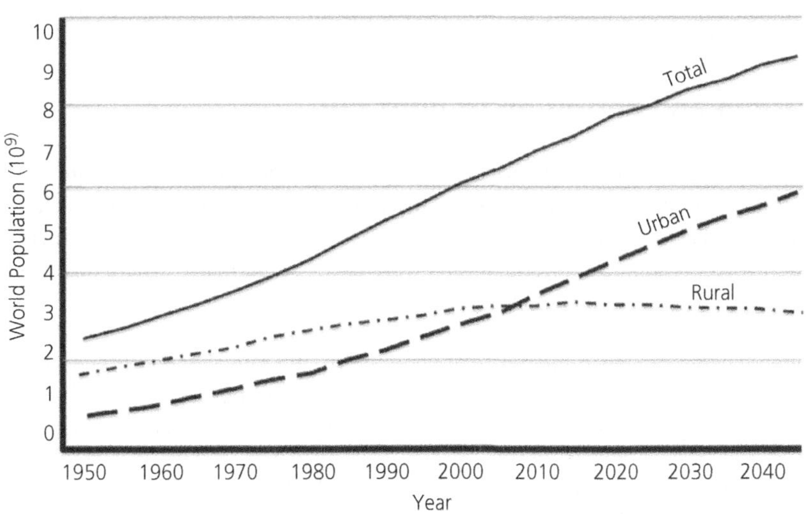

Source: United Nations data 2022

motorisation is not possible, even with a very large investment in a highway network and car parking. These are value negative activities. Indeed, increasing public spending on transport, especially roads, reduces resources for other important public services – for example, health care, education and housing.

A new approach to this is exemplified by the 'Strong Towns' movement (Marohn, 2020), which advocates a citizen-led investment, both time and money, to maintain and improve areas, and increase local value added. This approach has focused on the use of benefit-cost analysis, and shows this can distort the use of resources away from real economic value addition.

The alternative to private car use is public transport, which in many cities means buses. In developed countries, bus use has been declining for some years. Indeed, earlier studies (Webster, 1968, 1977) showed that, in the UK, each first household car reduced bus use on average by 390 trips annually, while the second car took away a further 250 trips annually.

In the US, motorisation is further advanced at a saturation level for many years with 90% of households having at least one car. This has not stopped an increase in car ownership, with multiple car households, and a level of 875 cars per 1000 population (Webster, 1977). This compares with the UK at 550. Bus use in Britain has been declining since 1955 by about 2% yearly. Most of the lost trips have switched to cars, creating traffic congestion. What alternative public transport can attract some car trips, and so reduce congestion?

While many cities have publicly owned public transport systems, most are bus based and for political reasons, subsidised. The fare income does not cover the operating costs. This is a burden on taxpayers, many of whom have their own car. There should be a way to break out of this cycle of declining bus usage and increasing subsidies. Light rail and tramways have been shown to attract trips from cars.

This book sets out how to design and operate self-funding systems that not only cover their operating costs but can contribute to repaying the investment capital. The UITP Light Rail Commission (UITP, 1991) showed that light rail/tramways are feasible in urban areas with populations less than 200 000, with a catchment population density of 1250 people per km^2. The critical factor is not the total settlement population but the population within the catchment area of possible tramway/light rail lines.

1.2. Origin and definition of light rail

Transportation systems using rails have been developed over more than 400 years. Initially, rails were installed in mines to make the removal of minerals easier when human haulage was employed, since hard wheels rolling on hard rails have a lower rolling resistance than on earth. Soon it was clear that rails could also provide guidance, making track systems practical in confined spaces. Initially, wooden wheels running on wooden rails were used. In Germany, where this was pioneered, the rails were called 'traams', meaning beams (the rails).

While wooden wheels can easily be replaced when worn, the rails were a different matter. At first the rails were overlain with strips of iron, significantly extending life. As metallurgy improved, iron rails were developed, giving an even longer life, although at a higher cost than wooden rails.

Two methods of guidance were developed. One employed 'L' shaped rails with plain wheels. The other used plain rails with flanges on the wheels, engaging the inside face of the rails. This latter was found to be better and has become the norm.

Rails, when coupled with steam locomotion, were used for more than local freight transport, becoming free standing companies, as railways/railroads. Operating between towns, they replaced both road and water transport for inland movements. It was realised that there was a demand for passenger movement between towns, as rail offered faster, more comfortable and cheaper transportation than horse-drawn coaches on rough roads. The first such railway was built and opened by George Stephenson in 1825 between Stockton and Darlington. The first 'modern' railway opened in 1830 between Liverpool and Manchester. Railways spread quickly around the world.

There is a record of a street railway in 1828 in Baltimore (Dunbar, 1967), but the birth of light rail can be dated to 1832 when Irishman John Stephenson (no relation to pioneers George and Robert Stephenson) realised that rails could revolutionise urban transport, previously dependent on walking or, for the more affluent, horse-drawn carriages (Figure 1.2). He laid rails in New York streets, using a large carriage with iron-flanged wheels pulled by horses, to start a new way of urban movement.

Horses can easily move 50 passengers in a carriage on rails, compared to only 15 on dirt roads, muddy in wet weather, or 8 t (17 600 lbs) of freight compared to 1 t (2200 lbs) otherwise.

The first rails were flat with an upstand of about 25 mm (1 in) on the outer edge (Figure 1.3). These thin plates were nailed directly onto longitudinal wooden beams, providing vertical strength and stiffness. The dirt road was then remade around the rails. The first European tramway (streetcar) system was opened in Birkenhead, England in 1860 by the American entrepreneur George Francis Train. Subsequently, three lines were opened in London in 1861. While private carriages travelling along the rails had a very smooth ride, crossing the rails was jerky and some carriage owners claimed their carriage wheels were damaged, so the rails were removed.

It was not until 1872 with the advent of grooved girder rails, providing a flush road surface, that new lines were opened (Figure 1.4).

Surrounding the rails with stone setts or wooden blocks provided an improvement in the road surface for all users but the tramway company had to pay for upkeep. The cost of keeping 12 horses for every tramcar in service was at least half of the ticket income, whereas staff pay was only about 10%. For many horse-powered tramways, the sale of horse manure equated to the annual profit. This gave an economic need to find a lower cost traction alternative. At first, modified steam locomotives were used in industrial towns. In less industrial towns, the smoke and noise were

Figure 1.2 Early horse drawn streetcar

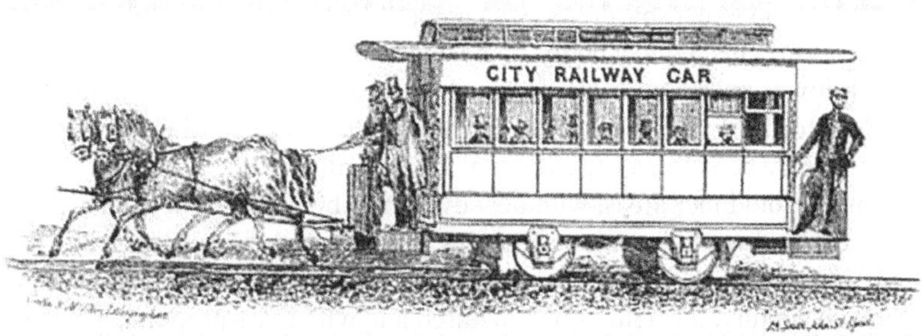

Figure 1.3 First UK streetcar track

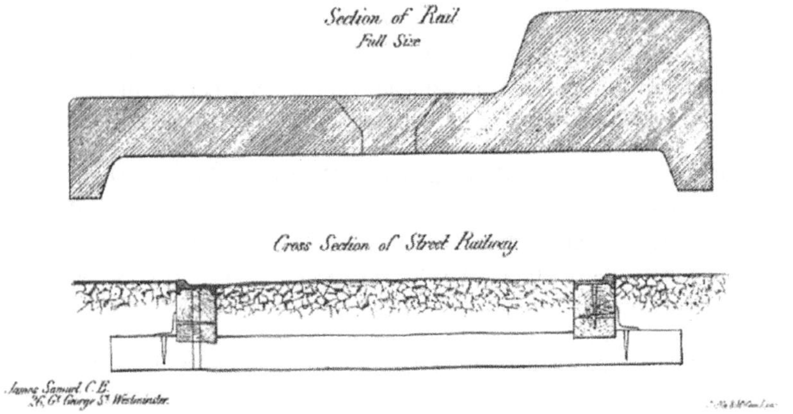

Figure 1.4 Grooved girder rails used for nineteenth-century tramways

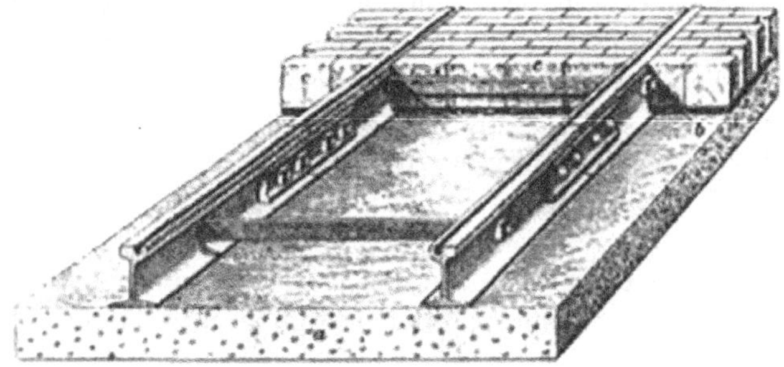

unacceptable. An underground cable haulage system, still used in San Fransisco and Llandudno, was employed in some places – for example, in Edinburgh until 1926. An endless cable was driven by a stationary steam engine and the cars had clamping grippers for movement.

The real breakthrough came with the application of electric power. Various methods were tried to provide the electricity to a moving car, including batteries. The overhead line, with a trolley pole pressing the underside, invented by Frank Sprague and first tried in Virginia, US, in 1885, soon became the world norm. This led to a boom in the construction and operation of electric tramways which, with much lower operating costs, lowered fares. What had been a middle-class mode of transport could be afforded by poorer people, becoming mass transit. Many municipal authorities not only took over the ownership and operation of local tramways but used them as a way of clearing insanitary houses from inner areas, with new dwellings being constructed in outer areas linked by tramways that enabled workers with new homes to travel to their employment.

During the twentieth century, the automobile matured and with mass production became an affordable means of transport for many people. This gave competition for tramways, some of which were modernised, while others switched to running 'cheaper' buses. This tram to bus conversion, however, in many places accelerated the purchase and use of private cars. Many cities and some countries became bus only, while many municipal authorities built more roads to accommodate growing levels of private car traffic to overcome traffic congestion.

By the end of the twentieth century and influenced by reports like 'Traffic in Towns' (Buchanan, 1963), cities realised that it would be financially impossible to accommodate cars being used for all travel in urban areas, even if there was space available. Los Angeles became the nearest to achieving this, with a freeway network of 850 km (528 miles) for a population of 3.9 million, with 88% of the population owning a car. Despite this highway provision, Los Angeles has all day congestion and a serious problem of toxic air pollution. Ironically, Los Angeles once had two urban rail operators – Red Lines and Yellow Lines – with a streetcar system using 2150 vehicles until 1963. Red Lines had a network length of 885 km (550 miles) and 900 vehicles, while Yellow Lines had 20 routes with a per capita ridership twice that of San Francisco.

Research shows that car users will switch some trips to a rail alternative, if available. While underground systems might be the ideal, the cost and time required for construction is prohibitive in many places. Added to this, motor traffic emits large volumes of toxic air pollution, affecting residents' health and straining medical services to treat illnesses caused by toxins, both gaseous and particulate. A search for a lower cost alternative in the US led to the reinvention of 'streetcars' as light rail.

Light rail is defined by UITP (1991) as a system of passenger transportation using rails and powered by electricity. The tracks can be laid in city streets, enjoying a high degree of priority to ensure reliable operations, including traffic-free streets offering good accessibility, especially to city centre destinations (Figure 1.5). Light rail can operate in the median of highways (Figure 1.6), at higher speeds and with a high level of reliability. Finally, light rail lines can also operate away from streets – for example, on an abandoned railway or road alignment (Figure 1.7) – to reach traffic generators or attractors more directly.

Figure 1.5 Example of system in pedestrian area, Freiburg

Source: Tony Young

Figure 1.6 Example of light rail in a highway median strip

Figure 1.7 Example of light rail on off road alignment

At the same time, light rail vehicles have been developed to provide smooth and rapid acceleration, carry more passengers, provide step-free access for those with disabilities, offer a high level of comfort and competitive travel times compared with automobiles (when including the need to park at the destination). Light rail operates safely in pedestrianised areas, bringing passengers to the heart of urban areas conveniently and quickly.

1.3. 'Trackless trams'

From time to time, companies promote tram-like public transport systems, the most recent being a group of Chinese companies. This is an articulated vehicle running on rubber tyres, electrically powered using batteries and periodically recharged at terminals. The present version is guided by magnets embedded in the road surface and detectors steering the vehicle. This concept was briefly promoted in Liverpool in 2022, and is presently under consideration near Perth in Australia.

This is a development of an earlier optically guided system (Figure 1.8), which is vulnerable to poor light conditions and weather obscuring the guide lines. Other guidance systems include buried electric cables, providing a continuous magnetic path, demonstrated in the UK in 1995 and promoted for lines in Liverpool and London. These were not taken forward, as the demonstration vehicle was prone to derail on curves when going only 10% over the required speed and crash barriers on curves raised the cost of the track.

There are variations of guided tram buses, including the Trambus service tried in Swansea and York (Figure 1.9), UK. Swansea City Council spent considerable sums on modifying streets on the

Figure 1.8 Chinese 'trackless tram' demonstration, white line guidance

Source: CRRC

Figure 1.9 Trambus in York 2006

Trambus route to make operation possible. These two projects used diesel powered buses, with all except the front steering wheels covered. These trambuses are driver steered.

In 2002, Caen in France opened a rail guided trambus Transport sur Voie Réservée (TRV), using vehicles provided by the manufacturer Bombardier (Figure 1.10). In 2011, after considerable difficulties, including vehicles derailing and crashing into traction poles and problems obtaining spare parts, the city council decided to scrap the TVR and replace it with a conventional tramway. The mayor commented that a more expensive tramway in the first instance would have been cheaper than TVR. Similarly, Nancy in France used the same system between 2000 and 2023, prior to it being scrapped.

Other forms of bus guidance have been developed, the most widely used being kerbs on either side of the track. This was first developed in Essen, Germany, to allow trolleybuses to operate through tram tunnels. Subsequently, kerb guidance has been widely used with a large system in Adelaide, where new tram lines are now being built, and in the UK, with the longest such system between Cambridge and St. Ives, mostly on the track bed of a former railway, operated with double deck buses.

All these systems have a common drawback. They cannot operate on flexible road surfaces (e.g. tarmac), since the continuous wheel impacts on a road pavement leads to fatigue failure, with the pavement changing from elastic to plastic, creating ruts, the ponding of surface water and subsequent surface disintegration. Guided buses all need rigid pavements (e.g. concrete) (Figure 1.11). This effectively means a special track is required, which offers little advantage or capital cost saving compared to light rail tracks.

Finally, guided buses have not managed to achieve significant modal shift from car travel, a point made in the US TRB Report 1221 (Tennyson, 1989) (see Section 1.4), whereas light rail and trams manage between 25% and 45% modal transfer from private car trips. The promoters of these

Figure 1.10 TVR in Caen showing the single guide rail, also used as an electrical return

Figure 1.11 Kerb guidance in Adelaide, Australia, and St. Ives, UK

schemes realise that buses have low quality perception in the urban travel marketplace compared to trams. Urban residents can see the difference and, while their behaviour in not switching from cars to trambuses is apparent, many are willing to make some journeys on real trams. This should be a strong marker for transport authorities seeking to reduce car traffic and increase the use of public transport. It is also another nudge to find ways to reduce the cost of building and operating real tramways.

1.4. The literature on light rail

In many developed countries rail enjoys a tiny share of the travel market. In the UK under 2% of trips and 7% of travel distance is by rail. Whereas roads carry about 95% of trips and 80% of travel distance. Surprisingly, railways have a devoted following far outweighing their use as a mode of transport or the market share enjoyed. Light rail is a part of the wider rail market. Most of the present literature on light rail is aimed at enthusiast readers. This is not to decry these publications, since many enthusiasts have encyclopaedic knowledge of light rail matters and offer useful insights into how light rail might be improved (e.g. Buckley, 1975; Dunbar, 1967; Garbutt, 1989; Hass-Klau 2004; Joyce, 1964; Powell, 1997; van der Bijl *et al.*, 2018). Identifying and applying this valuable information is another matter.

There is also a lot of material on the World Wide Web (e.g. APTA, Lightrail, Lightrailnow, LRTA, Transport 2000, UITP and Wikipedia). Much of this is ad hoc and difficult for a practitioner to apply systematically with confidence for the promotion, planning, design, implementation or operation of a new light rail or tramway system. This present book is aimed at those charged with designing, building and operating light rail systems, and is based on considerable practical experience, research and discussions with other knowledgeable practitioners.

The organisation representing public transport operators worldwide (UITP) has within it various special boards. One is for tramways and light rail systems. This has produced many reports, both about operational matters, as well as wider and more strategic concerns.

Of the books written by professionals for professionals, most come at the subject from a supply side perspective (Coffey and Kuchwalek, 1992; Lesley, 1991). In the 1920s in North America and

1930s in Europe, when car ownership was very low, such an approach made sense, since tramways, streetcars and other rail systems were operating for 'captive' riders. This is particularly true of those volumes which try to show why one mode of public transport or transit should be used in one circumstance but not in another (Tolley and Turton, 1995) (Table 1.1).

This is usually based on peak traffic (forecasts). The debate about bus, light rail and metro falls into this category, and is based on the fallacious assumption that potential passengers consider all these modes to be of equal merit and quality. Market research shows this clearly not the case (Fernandez, 2023; Lesley, 1974).

A supply side perspective is inappropriate in developed countries, as public transportation (bus and rail) is often a minority carrier of personal travel, with most trips made by car, even in major cities like Berlin, London, New York, Paris, Toronto and Washington DC. For the promoter of transit improvements to approach the subject with a supply-side perspective risks missing target markets and overestimating the potential patronage. The recent awareness of climate change and the need to reduce green house gas (GHG) emissions is, however, leading to policies that on the one hand are improving public transport and on the other are attracting trips from private cars or discouraging use. On the basis of the above analysis, the Manchester Metrolink, with a capacity of 2000 passenger capacity per hour, would never have been built. It was justified on the basis of an economic analysis.

The US Transportation Research Board Report No. 1221 (Tennyson, 1989) provides evidence for the promoters of light rail about the level of patronage that can be attracted. This was the result of analysing 40 years of transit data in North America. On the other hand, the UK National Audit Office Report on Light Rail (NAO, 2004) should make salutary reading for all those promoting better public transport, especially light rail, from the pitfalls of over optimistic patronage forecasts and cost over runs.

Similarly, the analysis of the 'Strong Towns' (Marohn, 2020) movement on the utility of public capital spending and the reliance on cost-benefit calculations sounds a note of caution. Financial liabilities are being created faster than the public purse can maintain, leading to the run down of other sectors – for example, the development of potholes in roads.

Table 1.1 Capacity of urban transport modes

Mode	Max. Capacity People/h	Av. Speed (km/h)	Stop Spacing (km)
Bus on road	9000	16	0.2
Bus/busway	20 000	26	0.8
LRT	40 000	26	0.5
Metro	60 000	32	1.5
Car on road	1000	19	–
Car/freeway	3000	72	–

From Tolley and Turton (1995)

This book partly redresses that imbalance by looking at market research on what is needed to get car commuters onto (light rail) transit, through attraction rather than punitive measures like parking restrictions and pricing, or congestion charging and other methods of real-time road user taxes. It also considers how to deliver light rail systems that can generate enough operating revenue to at least replace assets as they wear out, and possibly even service the investment cost. This means that subsidies are not needed, by grants from the public purse or taxpayers. Being able to depreciate assets for replacement is the key requirement for a financially sustainable investment. This perspective is especially important for those projects being promoted by public bodies, with the capital cost from taxpayers, and the revenue not exceeding the operating cost by enough to replace assets as they wear out.

This theme was the subject of a serious paper comparing private and public tramways (Yearsley, 1993). Yearsley compared private tramway companies in London with those of the municipal London County Council (LCC). LCC tramways in years when fare surplus were generated were paid into the general accounts and reduced the need for local taxes. On the other hand, when new equipment or track replacement was needed, politicians treated this as a drain on public funds. Yearsley showed that had the LCC run its tramway as a private company, all equipment replacements would have been covered out of depreciation and, on average, an annual 4% dividend would have been paid to the council.

1.5. The development of light rail

The transition from streetcar/tramway to light rail began in the 1970s, when the realisation began to emerge of the growth of private car use and the limitations of buses. Even in a command economy city, like Budapest in the 1980s, the performance of slow on-street tramways delivered quicker door to door journeys than faster metros (subways) (Figure 1.12) for journeys up to 8 km (5 miles).

Figure 1.12 Journey time comparisons in Budapest 1985

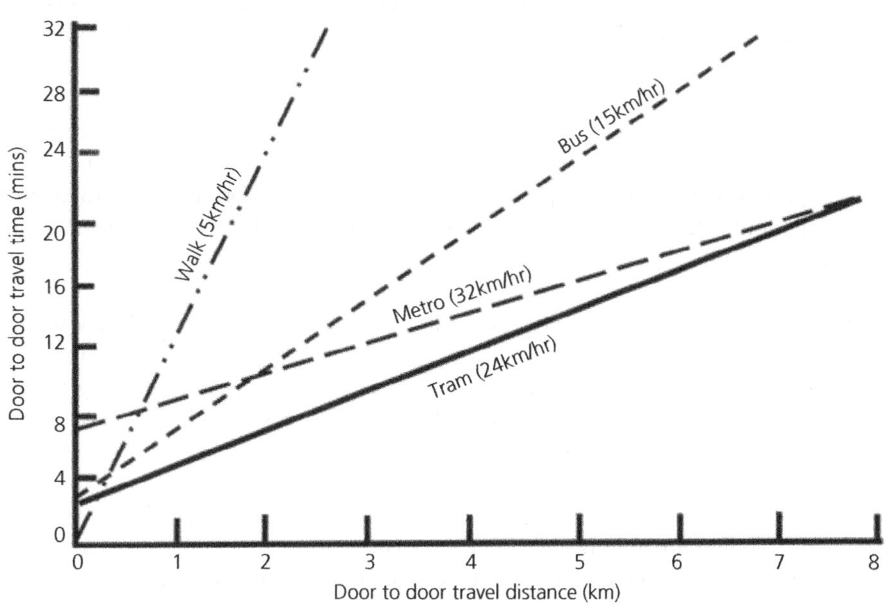

This was due to better accessibility to stations and the short waiting times for frequent tram services. Arguments for such systems began to be aired in places with high levels of car ownership (Western Europe). Before the fall of the Berlin Wall, Gothenburg-style zoned city centres were being introduced. These reduced car traffic (through traffic) and protected tram tracks to improve operating speeds (Lesley, 1986).

By the 1970s, North American cities also realised the truth of Professor Sir Colin Buchanan's seminal report 'Traffic in Towns' (Buchanan, 1963), advising the UK Government on future policy for urban transport needs. He showed that the latent demand for private car trips was so high, even in low density cities like Los Angeles, that there could never be enough road or parking space to meet peak travel flows entirely by car. He advocated a twin approach of better public transport and steps to manage peak car traffic to levels that could be economically, environmentally and socially accommodated on the existing road network. With North American cities reaching gridlock and downtown areas dying for want of acceptable access that bus systems did not satisfactorily provide, municipal authorities saw that rail transit attracted car commuters but could not afford to build new subways or metros.

Metro or subway systems are expensive and take a long time to build and open. The reasons for this relates to the need to acquire large sites for stations and the high cost of tunnelling. Pierre Laconte, then Secretary General of UITP, observed that light rail offers 90% of the benefits of a subway at 10% of the capital cost (Laconte, 1978). In 1860 this comparison was made by George Francis Train, 'a street railway can be constructed in as many months as an underground will take years. The former costs 3000 per mile, the latter 300 000' (Collins, 2001). Light rail or modern streetcars were seen as a quick and affordable solution by transport planners. The image of streetcars in North America was considered to be antiquated from historic slow services. The generic name 'light rail' was invented to re-launch streetcar transit. US cities used European equipment and approaches to build new low-cost light rail lines. The first to open was in San Diego, (1981) using German articulated vehicles. The system now has three lines with a total length of 109 km (68 miles) and 62 stations. In 2023, it carried over 30 million passenger journeys, the highest ridership of any US light rail system. Other systems quickly followed in Calgary and Edmonton also using German articulated vehicles.

By the 1980s policies of privatisation and deregulation were taking the stage politically in many countries. The aim was to remove the perception of public sector inefficiency and pork barrel politics. The rights or wrongs of either public or private sector as the promoter or provider of light rail systems will not be debated here. This book tries to show how light rail can be made to work through either approach.

The use of open competition is a cornerstone of the procurement policy for publicly funded projects to improve and reduce the costs of transit services all over the world. In North America transit remains highly regulated and largely publicly owned. There is also a debate going on over the difficulty of introducing innovation in light rail, when risk-averse authorities only seek equipment that is tried and tested, even if inefficient.

There is also the idea that tendering can transfer risks from public bodies to private contractors. Contractors bidding for light rail projects can easily budget for technical risks. The biggest risk, however, is the public client changing policies or plans. Thus, private contractors include significant margins to cover such risks. This raises bid costs. Contractors bidding for such projects have a high cost in preparing bids and a low chance of winning the contract. Factoring this further risk also increases costs. Thus, in many places there are now too few contractors willing to bid for light rail projects.

1.6. Environmental impacts

Transport consumes resources and impacts the environment. These can be divided into

- resource extraction to construct infrastructure
- resource extraction to provide vehicles and the motive power
- air emissions from power supply and operations
- noise and vibration
- community division.

1.6.1 Resource extraction

Raw materials are mined, refined and manufactured to create materials for vehicles, ways, maintenance and garaging. At the end of their working lives, these must be disposed of. Some materials can be recycled, others are disposed of. Both activities have environmental impacts, consume energy and release emissions into the environment. Fortunately, the size of this resource depletion is small in comparison to the environmental impacts of operating transport systems. The metals used in vehicles and tracks can be recycled.

1.6.2 Energy consumption and pollution

Transport is an energy intensive operation and, until recently, very oil dependent. In the developed world, over 90% of transport system operations depend on oil. Various commentators have focused on climate change and the need to reduce the emission of GHGs. This may also coincide with peak oil, when worldwide production will begin to decline, or pressures increase to stop oil extraction. In many countries, transport releases approaching half of all GHGs, as other sectors of the economy decarbonise. As well as this, many transport modes also release toxic pollution which damages the health and welfare of people. These also place a large burden on health services in treating pollution-based diseases. For environmental and health protection, new sources of motive energy are required that do not emit GHGs. In particular, the replacement of hydrocarbon fuels like oil is a high priority so that carbon emissions can be reduced and climate change addressed.

The largest impact of oil consumption, however, comes from the impact of burning fossil fuels inefficiently. Over 75% of the energy contained in oil does not produce movement when used in internal combustion engines but is wasted as heat or unburnt fuel. Transport is the most energy inefficient sector of the economy. Electricity generation from hydrocarbon fuels only wastes about 60% of input energy and is rapidly converting to renewable generation. Transport fuels (oil) have been cheap enough to be used wastefully, because the price paid reflects the cost of extraction, rather than replacement or compensating for the environmental and health damages.

Rail systems are energy efficient compared to road transport, giving at least four times the passenger km compared to private cars on roads per unit of energy employed. Rail transport, including light rail, can be easily electrified and, when powered by renewable generation, is GHG neutral. The technology for electrifying rail systems is mature and there are several suppliers, so the costs are affordable. Added to this, light rail and tramways attract trips that would otherwise be made by private car, which then has an indirect benefit of reducing both GHG and toxic gas emissions.

Hydrogen (H_2) has been advocated by some as a replacement for hydro-carbon fuels. H_2 used in fuel cells to generate electricity, only produces water vapour. In making the H_2 needed, however, whether from processing methane gas (a fossil fuel) or using off peak electricity to split water into hydrogen and oxygen, the total fuel cycle efficiency is no better than hydrocarbons in internal combustion engines, and presently much more expensive.

Finally, health and globally threatening pollution released by transport systems is primarily from road transport. Oil is a hydro-carbon fuel, so burning it in oxygen produces carbon dioxide (CO_2). Approaching 50% of all CO_2 in the developed world now comes from transport. The actual amount and the proportion of the total released by transport is growing, as other sectors of the economy adopt efficiency measures and renewable energy sources. CO_2 has been linked to climate change, which is fundamentally altering the habitat of the world, making the survival of some fauna and flora problematic in many places, especially if sea levels rise, flooding coastal areas and fertile land.

Waste emissions from transport operations also include carbon monoxide, oxides of nitrogen and micro particles (PM10). All of these are associated with several cardio-vascular and lung diseases. Most recently researchers have linked high levels of PM10 and smaller particles with higher risks of women miscarrying babies and dementia among the elderly.

While there is public concern over deaths and injuries caused by transport crashes, many more people die or are made ill by transport pollution (Figure 1.13). In the UK under 2000 people a year are killed in road crashes. In comparison, about 60 000 die from transport pollution diseases.

Figure 1.13 Graph of air pollution on a typical UK urban road

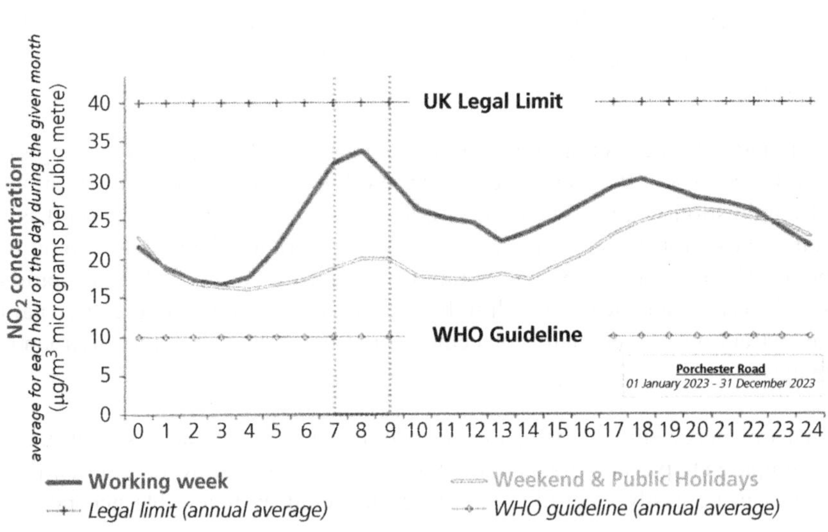

At a local level, transport emissions reacting with sunlight causes petro-chemical smog, making the health impacts much worse. Los Angeles is notorious for this because of a combination of topology and climate. Inversion smogs occur in many cities. The European Commission has issued air quality directives, setting the maximum concentrations of various combustion products. Similarly, the World Health Organisation has produced what are considered to be safe levels of urban air pollution, especially where children are exposed. It has been recorded in many cities that children living near busy roads suffer much higher levels of respiratory diseases, including asthma, than children living in the suburbs or in rural areas. Most British cities fail to meet the WHO Air Quality Standards.

1.6.3 Noise and vibration

Road traffic noise is the most widely reported noise nuisance in most cities. The WHO recommends a maximum daytime level of noise of 55 dB(A) for social living, and a night-time level of 45 dB(A) (Figure 1.14). Many city streets have noise levels of 70 dB(A) and above, which is twice as loud, making social conversation difficult, if not impossible. All vehicles contribute to this noise. For most motor cars, the significant noise generator is now the tyres not the engine, as a result of improved engine and silencer/muffler design. Larger vehicles are noisier with diesel engines being over 70 dB(A). Better engines and transmissions will not reduce the overall noise level below 70 dB(A), since that is the noise generated by rubber tyres on the road pavement.

Roads are not perfectly flat, so vehicles bounce as they progress. This bouncing is amplified by the load of the vehicle and nature of the suspension. The bouncing is carried by the road pavement into adjacent buildings, where windows and fittings can rattle. There is no record of buildings being structurally damaged by such vibrations but they can be annoying, and occasionally ornaments fall over and break. Light rail systems are much quieter, generally about 60 dB(A).

Figure 1.14 Noise of motor vehicles at different speeds

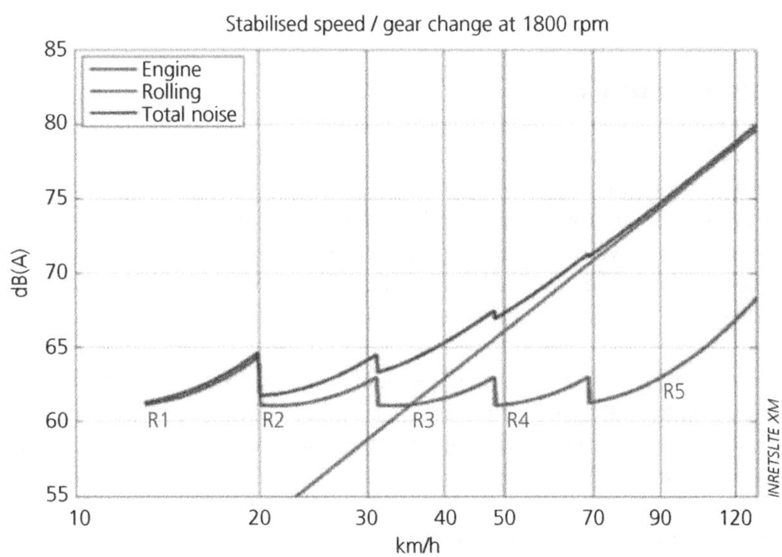

1.6.4 Community division

'The wrong side of the tracks' is a well-known social phenomenon, where different income groups occupy areas separated by railroad tracks in American cities. Highways can also divide communities. As early as 1963, an environmental traffic flow level was defined (Buchanan, 1963) for residential areas as under 200 vehicles per hour. Subsequent research shows that how pedestrians cross roads is dependent upon the volume of vehicle flow. The higher the flow, the more likely pedestrians will cross at right angles. Put the other way round, if pedestrians cross the road at obtuse angles, the traffic is likely to be near environmentally acceptable level.

Finally, many studies have shown that the number of neighbours people know on the other side of the road is highly correlated with traffic flow. The higher the traffic flow, the smaller the social network, with few people from the other side of the road being connected. Of course, some roads have been built (e.g. freeways and motorways) to exclude pedestrians and to make crossing the road difficult or impossible. These can be even worse in separating communities.

There are however places where major roads have been converted to inclusive community use. One example was the Embarcardero Freeway, built in 1968 in San Fransisco but collapsed by an earthquake in 1989. This has been converted into a linear park. A similar thing happened in Valencia, but the earthquake was political in that an alignment reserved for a freeway through the heart of the city became a linear park, walk and cycleways. Both cases, and other similar cases, reduced the volume of motor traffic and with it noise and air pollution. They also increased the health and wellbeing of local people who are now able to take exercise in the open air.

1.6.5 Light rail impacts

Light rail lines, offering the same capacity as a six-lane highway with car traffic, can operate at up to 1 per minute intervals, so makes social cohesion easier to achieve (Figure 3.10). Being electrically operated, no air pollutants are emitted in the street and, if powered by renewable generation, will be CO_2 free, needing no fossil fuel and so is environmentally and energy sustainable. Electric traction and light rail vehicles are quiet, typically emitting less than 60 dBA (Figure 1.15), and steel wheels on steel rails produce little in the way of ground-borne vibrations and provide smooth riding for passengers.

1.7. Planning light rail

This book will not consider in detail the legal requirements found in different countries, or even cities, for light rail lines to gain approval to be built and operated, as there are many approaches. Instead, detailed consideration will be given to the scientific processes by which routes can be identified and evaluated; environmentally and economically. These processes can be applied to most urban areas, irrespective of the legal approval requirements. They are based on established demographic, economic and geographic analyses that allow potential catchment areas to be compared for the trip generation and the attraction of new light rail lines.

Similarly, the way in which systems can be operated will be evaluated, to provide an acceptable quality, by identifying the optimum use of resources and considering how passengers react to different combinations of service provision. These aspects of planning are universally accepted and the principles well documented (e.g. Banister, 2002; Bruton, 1970; Faludi, 2004; O'Flaherty, 1997; Pharoah and Apel, 1995; Starkie, 1976; Tolley and Turton, 1995; Vickerman, 1991; White, 2005). Use is made of generalised costs as a proxy for the complex way in which people compare different travel options, and then selecting one for making a particular journey, or in choosing home location

Figure 1.15 Passing noise level from City Class trams

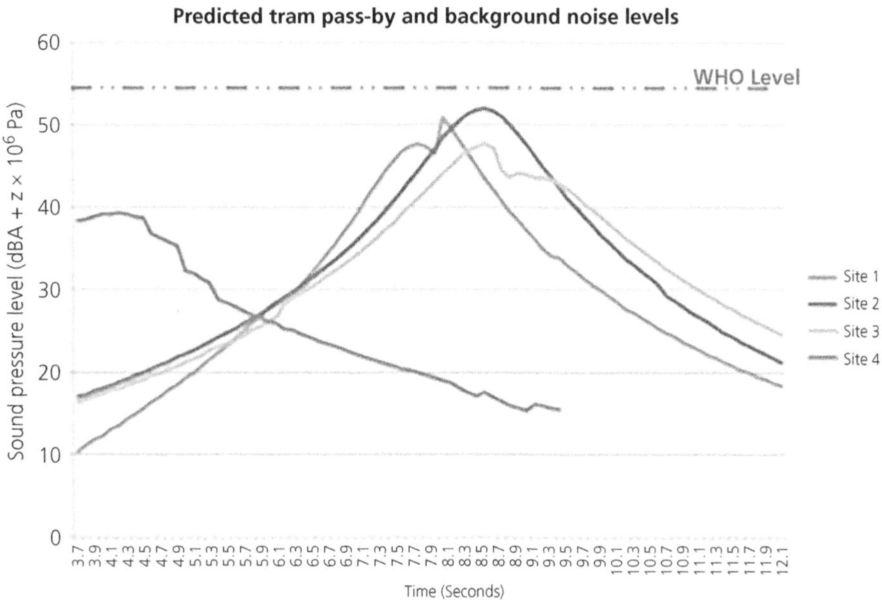

Source: The author

to minimise travel cost or the time required for the most important travel (the journey to work) if these are remote and not home based.

1.8. Engineering light rail

The physical forces created by the movement of (light rail) vehicles, the capacity of the road pavement and ground to carry loads, and the strength of materials and structures are all universal principles of physics and civil engineering. The detail of how to determine the best designs and material configurations will be examined, together with different technologies and installations, to achieve a safe operating environment. These principles are transferrable from one place to another, allowing light rail engineers to use previously successful experience when designing a new system and therefore reducing the costs and risks of new installations. This is the fundamental scientific approach by which humanity has adopted technological discoveries and inventions, and improved the health and wealth of the population.

Safe, reliable and robust light rail systems that are economic should be the objective for all promoters. This book provides a systematic basis for achieving that, and, in doing so, looks at attempts that have not always been successful, thus helping new light rail designers to avoid costly mistakes. It also draws on widely available texts on general engineering aspects of highways and railways (e.g. Atkins, 1980; Morgan, 1973; Profillidis, 2000; Salkield, 1953; Sharp, 1970; Slinn *et al.*, 1998)

1.9. Economics and commercial environment

When light rail schemes are being promoted by public authorities, the capital required usually comes from a budget provided by taxpayers. This is not limitless. Demands on taxes for public capital works and revenue payments, like transit subsidies, usually outstrip taxpayers' willingness to fund. Inevitably, a political mechanism is needed to justify the use of 'free' money, especially if operating at subsidised fares (Lesley, 1987). Benefit-cost analysis was developed as one way to demonstrate and justify the non-financial benefits of such public investments, and to identify the 'opportunity cost' compared to other potential public investments – for example, hospitals, schools and so on.

Transit benefits include environmental improvements, reducing congestion, encouraging economic activity, and reducing the need to expand the road network and so on. Politicians also undertake an electoral calculus about projects that will win the most votes in different places. In this scenario light rail often has a low priority compared to schools, hospitals and even highways on which voters can drive their cars. Even in cities with widespread traffic congestion or high levels of air pollution, there has not been enough pressure to provide light rail for more than a token diversion of car trips. Rarely is highway capacity removed in such cases. This means that suppressed car trips quickly fill the space vacated by trips attracted to light rail. Thus, no overall environmental improvement can be identified.

There are texts by transport economists that can enlighten this debate (e.g. Glaister, 1981; Lesley, 1996; Townsend, 1969). Added to this is the analysis of the 'Strong Towns' movement (Marohn, 2020), which demonstrates that continuing public capital investment does not expand the tax base of urban areas by enough to cover even the maintenance cost of all public plant, let alone replacement when due. 'Strong Towns' argues that using cost-benefit analysis to justify public investments distorts the use of scarce public funds and that the basis should be financial viability; new investments should pay for themselves from the revenues and taxes generated. This can come directly (e.g. fares) or indirectly (increases in local tax income).

To improve traffic speed, the geographer Mogridge (1990) showed that this can be possible by raising the speed of transit services, as this attracts marginal car trips. With less road traffic, the speed of remaining vehicles increases.

In North America light rail systems, capital costs have varied widely between the low achieved in San Diego ($10 million/mile = €5m/km of first line) (Figure 1.16) for a 100% surface level system,

Figure 1.16 San Diego (an affordable light rail system) – typical stop

Source: San Diego Transit

and the high in Buffalo ($100 million/mile = €43 million/km) with 80% underground 8.4 km, with the middle near the European average (€18 million/km).

At the UK Parliamentary Light Rail Inquiry held in 2009, the question of light rail costs was raised by the Chairman, Paul Rowan MP. New light rail construction costs in the UK are over £20 million per km ($50 million/mile) and rising. Andrew Braddock (formerly Transport for London) reported that the target price of new tramways in France is €14 million/km (c. $33 million/mile). A recent extension to the Brussels tramway cost €8 million/km ($19 million/mile). Karlsruhe on street track extensions are €7–8 million/km ($16–$19 million/mile). A senior German tramway director in Stuttgart commented on the French costs: 'The French figures often includes restructuring/reshaping of the entire street environment. Thus, I do not consider them apt for cost comparison of pure permanent way construction. Our figures for a double-track alignment (including overhead, stops and signalling) vary from €5 to €10 million per km' ($12–24 million/mile).

In spite of German construction workers being better paid than British, German tramway construction costs are less than half those in the UK. This may explain why there are more tramways and extensions being built in Germany than the UK. The proposed GLUAS scheme in Galway at €9.5 million/km ($22 million/mile) is consistent with German tramway costs. For light rail to play a larger part in providing an energy sustainable alternative to urban car trips, construction costs need to be reduced. Ways to do this are discussed in this volume.

Light rail schemes that do not cover their operating costs will need subsidies. This assumes that the capital is a free grant and not a loan that has to be repaid. For politicians funding and subsidising light rail, this can seem expensive, especially when motorists driving cars are an important source of tax revenue, as well as potential electoral support. For public finances, therefore, light rail can seem to be a negative investment, needing taxpayers' money to build and operate, and reducing the tax take from reduced fuel sales because of trips attracted to light rail transit (LRT).

Many US systems have been funded by bond issues underwritten by local sales tax (increases), typically about 2%. These tax increases are usually approved by a referendum of electors. A study by the University of Texas at Dallas asked electors in Dallas why they had voted to increase sales taxes to fund a new rail project (DART). The answer was that most electors did not necessarily see themselves as transit users. Enough other car drivers however would be attracted to the light rail system, to reduce traffic congestion and make driving easier for the rest (Figure 1.17).

The advent of low-cost airlines shows that the public sector model for transport is not the only one that provides an economic or acceptable service for passengers. In places where transit is operated commercially, there is an argument that new light rail systems might be promoted by private companies and financed privately based on generating operating surplus from fares. Looking at the economics of light rail from a commercial perspective could help to unlock funding for new schemes that are financially viable that otherwise could not be afforded by the public sector. Can privately funded light rail schemes also meet municipal aspirations and increase civic pride? To achieve this there needs to be a genuine partnership, which is usual in other commercial and real estate developments.

In this case, the potential profitability of new schemes and the dividend that will be paid to investors will determine if a privately funded scheme can be successful. The likely viable route(s) will be through areas that are economically active, with high levels of mobility and congestion, rather

Figure 1.17 Dallas area rapid transit stop

Source: Dallas Express

than areas needing regeneration, if a scheme is to be attractive to private investors. Once a base network has been opened, extensions can be considered on a marginal basis, not needing such high internal rates of return (IRR) as the first lines, by sharing resources – for example, the depot, vehicle spares and so on.

In summary, an affordable light rail system is one that covers the operating costs from revenue, can depreciate and replace assets as they wear out, and should the capital investment be provided by the private sector, service the investment debt. To achieve this the system and installation should be the minimum, commensurate with safe and attractive operations. This means

- few, if any, special structures or tunnels, bridges or viaducts
- capacity to meet the near-term operating needs only but capable of expansion
- alignment and stations to maximise patronage and revenue
- alignment and traffic priority to minimise journey time and fleet size.

Achieving these also requires considerable engineering and operating inputs, as well as accommodations to minimise the negative urban impacts. Such impacts occur while the light rail scheme is being built and permanently once it is open. Changes in political leadership can have a big impact on the successful promotion of light rail schemes. Politicians with a green agenda can accelerate the construction of new light rail lines. This should help reduce dependency on oil and cars for urban trips. Similarly, legislation can force car builders to improve efficiency and sustainability, as in the UK only new battery cars can be sold after 2035.

Other developers are working on new cars to give more miles per energy input, reducing carbon dioxide and other emissions. These, however, do not address the fundamental issue of too many vehicles using a finite-capacity urban road network, nor the negative impact from trying to accommodate more car traffic. This is a major issue of the 'Strong Towns' movement in the US, and its successes, using a 'bottom up' approach with modest interventions, to make urban living more civilised.

1.10. Local and government support

1.10.1 Why?

Section 1.10 discusses the importance of liaising with and getting co-operation from public authorities when promoting a new light rail or tramway scheme. This is critical when public authorities are gate keepers of access to local highways or land that can be used as a depot. Clearly if a public authority is itself promoting the scheme, it will seek the support and co-operation of the private sector in the form of the Chamber of Commerce, local construction companies and property developers.

1.10.2 US and others

In the US there are several anti-light rail lobbyists, like Wendell Cox, Randal O'Toole and Bill Vincent (of the Breakthrough Technologies Institute), advocating bus rapid transit (BRT) as a cheaper option than light rail. In the UK, BRT has also been funded by central government instead of light rail, with schemes in Cambridge, St. Ives, Luton, Leeds, Dunstable, Manchester, Bristol and Crawley.

Even serious transport commentators sometimes criticise light rail as being expensive, and that buses (on busways) could do the same job for much less money. This was a policy followed in Houston in the 1980s and 1990s with the construction of over 100 miles of high occupancy vehicle (HOV) lanes and roads on six many major routes. This approach fails to differentiate between supply side and demand side considerations.

More importantly, car commuters do not perceive buses to be the same as rail, and this is an important market failure by those promoting improved public transport. While there have been ad hoc studies in different cities on the market assessment by residents, and especially car commuters for their views of alternative forms of transit, none has been as systematic and evidence based as that published by the US Transportation Research Board Report No. 1221 (1989).

Report 1221, entitled 'Impact on Transit Patronage of Cessation or Inauguration of Rail Service', covered a 40-year period. It studied the impact of closing (light) rail lines on travel patterns, as streetcar and commuter rail lines were abandoned and replaced by buses. In many cases, buses were later abandoned due to the growth of car usage. Towards the end of the study period, the impact of opening new light rail lines, to replace buses, on travel patterns was also studied. The closure of rail lines, replaced by buses, often with enhanced service frequencies, resulted in a loss of transit patronage of between 30 and 40%. The passengers lost to transit transferred to car travel.

Considering the opening of new rail lines, where a comparison could be made with former bus lines, or where bus lines were replaced by light rail, Report 1221 found that significant rail traffic was attracted from car trips. On a like for like basis, rail carried about 40% more passengers than bus lines (within a range of 35%–45%). This extra patronage was attracted from car users. This figure is higher than European experience because of the higher level of car ownership and usage in the USA. The TRB Report concluded: 'In most cities served by buses exclusively, transit riding has declined 75 per cent over the past 40 years. Exclusive busways have not made much difference absolutely, but they have helped relatively. In 11 areas with updated rail transit facilities, ridership has increased markedly, often by more than 100 per cent. In two of these areas, the transit systems are attracting more ridership than they did when gasoline and tires were rationed. It appears that rail transit makes a great difference in ridership attraction, with attendant benefits.'

The report continued: 'When these service conditions are equal, it is evident that rail transit is likely to attract from 34 per cent to 43 per cent more riders than will an equivalent bus service. The data do not provide explanations for this phenomenon, but other studies and reports suggest that the clearly identifiable rail route; delineated stops that are often protected; more stable, safer, and more comfortable vehicles; freedom from fumes and excessive noise; and more generous vehicle dimensions may all be factors … Those engaged in the alternatives analyses (*of transit modes*) and similar studies would be well advised to consider these differential factors before making service recommendations or traffic relief assumptions. Future problems with air pollution, congestion, and funding may all be seriously affected by these considerations.'

While there has been no similar comprehensive study in other countries, trams (streetcars) were abandoned in Britain during the period 1930–1962, all replaced by buses. In most cases the new buses were more comfortable than the 50-year old trams they replaced and ran faster and sometimes more frequently. In all cases the patronage declined immediately by between 30% and 40%, and continued to decline at an average of 2% per annum and overall by 70% (Figure 1.18) between 1955 and 2020.

France also abandoned most urban tramways in favour of buses from 1950 onwards. Since 1985 when the new Nantes system opened (Figure 1.19), some 30 new light rail systems have been opened. Nearly all these replaced the busiest city bus lines. On opening the new light rail lines, the patronage jumped within a year by a minimum of 30%, compared to the previous bus services. These new tramways in France resulted in a reorganising of bus services to operate as feeders. As a result, bus use has also increased. In Britain, seven new light rail lines have opened since 1992. Some of these have used old or closed urban rail lines. Market research of the patronage of these lines has shown that between 20% and 40%, with a national average of 25%, of passengers transferred from car commuting.

These data indicate that there are qualitative differences between light rail and bus services, which heavily influence customers, and, therefore, ridership and revenue. In particular, light rail has

Figure 1.18 UK bus use

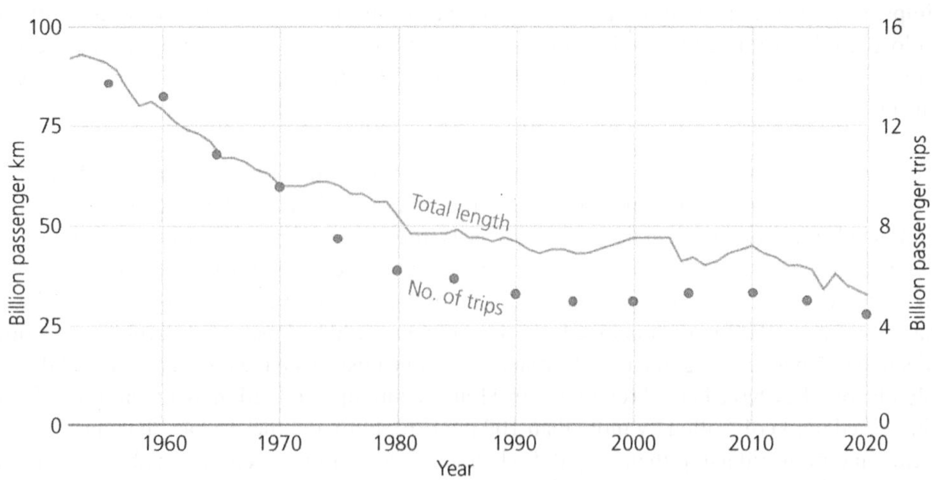

Figure 1.19 Nantes. France's first new light rail stop

Figure 1.20 Houston area rapid transit in street stop

shown itself able to attract significant traffic from private car commuting. This modal transfer is critical to help cities address and reduce traffic congestion, air pollution and other environmental problems. Houston opened its first light rail line in 2004 (Figure 1.20). This quickly attracted patronage, so that at the end of the first year it was carrying 10% of all transit trips in the metropolitan area. Construction began in June 2008 on five more LRT lines for a 30 mile-long light rail network. Meanwhile, HOV/bus lanes are being tolled for single occupant car use, implying the failure of bus services using them.

1.10.3 Government studies in the UK

Experience in the UK can also provide some lessons for those around the world promoting new light rail systems. Since 1990, new tramways have opened in Birmingham, Croydon, Edinburgh, Manchester, Nottingham and Sheffield. These have experienced mixed fortunes. The first in Manchester (1991) (Figure 1.21) linked two existing suburban rail lines with city centre street running. This quickly doubled the patronage of the former rail lines and was soon generating a surplus of £5 million per annum. Subsequently, extensions have been added linking Eccles, the Airport,

Figure 1.21 Manchester Metrolink, enlarged St. Peter's Square junction

Rochdale and Ashton under Lyme. A second city centre link helps to distribute traffic and reduce potential congestion. This involved a rebuilding and considerable expansion with a junction at St. Peter's Square, doubling the number of platforms at the stop.

The next to open, Sheffield (1995), was a larger network of three lines but quickly lost money with revenues less than operating costs. When planned, the tramway was expected to be part of an integrated service with local buses acting as feeders. Central government, however, deregulated bus services in 1986 (HMG Transport Act 1985), so the tramway faced unexpected competition from privatised local bus companies. Two of the lines are virtually all in street (Figure 1.22), and one to Meadowhall reuses an old railway alignment. The transport authority invited private operators to take it over on a 25-year concession. Stagecoach Buses won the contract for a reported price of £1 million, and by astute changes to operating practice and marketing, reduced operating costs and increased revenue to the stage of making a surplus of £1 million per annum.

In 1999, a new tramway opened between Birmingham and Wolverhampton with a track gauge of 1435 mm, unlike the original tramway built with a 1066 mm gauge. It has short sections on street in the centre of Birmingham and in Wolverhampton, but most of the line reuses an abandoned railway through the derelict industrial area known as the 'Black Country'. This resulted in stations with poor accessibility, serving areas with little or no activity. One of the objectives of the tramway was

Figure 1.22 Sheffield Supertramway in City Centre

to regenerate this area. On that basis, an annual patronage of about 10 million was predicted. In practice, regeneration was much slower than expected and the annual patronage was only 5 million. This led to an operating loss which was absorbed by the local bus company operator. Since then, the line in Birmingham has been extended on street past the conference centre and symphony hall to Five Ways junction (Figure 1.23).

The Tramlink in Croydon opened in 2000. This was an interesting hybrid with Croydon Council as one of the promoters with a private consortium of constructor, equipper and operator raising half the investment capital required. Most of this network of three lines uses railway alignments, with street running links through Croydon centre (Figure 1.24). Initially town centre traders opposed

Figure 1.23 Extension near New Street Station

Figure 1.24 Croydon Tramlink through a pedestrian area

the scheme but quickly changed their minds when footfall and turnover increased, aiding their businesses. Almost from the start, Tramlink was popular, with traffic and revenue growing, giving the consortium a profitable return on their investment. Transport for London (TfL) made an offer in 2008, reported to be about £100 million, to buy out the consortium and take over the operation. This resulted in not having to compensate the consortium by £4 million per annum due to imposing low fare levels.

Three tramway schemes, in Leeds, Liverpool and South Hampshire, had their funding withdrawn by central government in 2004 and were abandoned.

Nottingham Express Transit opened two lines north of the River Trent (Figure 1.25) from the Railway Station to Hucknall and Phoenix Park, with park and ride stops near the terminals. Much of these lines share the same on street alignment, with a junction near David Lane serving the two termini. The system nearly doubled in size when two lines south of the River Trent were opened in 2014 to serve Beeston and Clifton. Unlike the schemes above, this was financed entirely locally by both Nottingham city and county councils, partly underwritten by a workplace parking levy.

2014 also saw the opening of the new tramway in Edinburgh (Figure 1.26) between the airport and city centre. This is described in detail as one of the case studies in Chapter 7, including the significant cost and time over runs.

Figure 1.25 Market Street stop, Nottingham

Figure 1.26 Edinburgh Tramway on Leith Walk

For various reasons, most of these schemes were subjected to government review after opening. The National Audit Office (NAO), which monitors all spending by central government, published a report (NAO, 2004) on an investigation into the light rail systems which had been opened between 1990 and 2004. This report showed that those schemes had been subject to cost over runs and construction delays. Furthermore, forecasts of patronage had been over optimistic and operating subsidies were required. Finally, many of the anticipated objectives (e.g. regeneration) had not been realised. As a result, further new schemes were not funded by central government. Indeed, those first schemes were seen as poor value for money when compared to similar schemes in continental European countries, which had capital costs about 50% of those in the UK, when delivering the same transport objectives.

The NAO analysis underplayed the way light rail can achieve other objectives – for example, environmental, economic, traffic or social. Many of these objectives, however, result in no increase in operating revenue or tax income. The NAO showed that, in the UK, light rail schemes rarely fully achieved all the objectives that were claimed by promoters. Indeed, the use of benefit-cost analysis, which tried to equate these disparate and often subjective elements, rarely shows light rail to be good value for money compared to other transport projects also aiming for environmental benefits, such as cycleways. Perhaps this is because traveller time savings from light rail systems are often small, or negative compared to car trips, and that car-users' time is assigned a higher value than public transport passengers.

As a result of the NAO report, the UK Government withdrew grant aid for planned light rail schemes in Leeds, Liverpool, London and South Hampshire (Portsmouth), and has forced the promoters of new extensions in Birmingham and Manchester to compete for funding from a transport innovation fund or use local resources. This is meant to support all capital projects throughout England. Both cities were seeking to use 100% of the yearly allocation, which even if spread over 4 or 5 years, would significantly reduce the funding for other transport projects in the rest of England. Scotland and Wales have separate capital funding schemes. A referendum in Manchester in December 2008 voted 79% against a congestion tax to pay for light rail system extensions, costed at £3 billion. Recently, Manchester has been awarded £1.5 billion, spread over 5 years, for light rail extensions and other projects.

Separately in Scotland, the Edinburgh tram project has been subject to two investigations. The first report (Audit Scotland, Interim Report, 2011) reviewed a scheme only half built, four years late, at a cost of £770 million by 2016 and carrying a third of expected passengers. This concluded that confusion between the client (Edinburgh City Council) and the contractors contributed to the problems reported. The second report published in 2023 (Hardie) was a public inquiry chaired by Lord Hardie. This sought to discover the reasons for these problems. His conclusion was that the client did not have the required expertise or experience in house to be able to specify or supervise a new tramway. During this inquiry, further funding was raised and the line was extended to the intended northern terminal, Newhaven, which opened in 2023 at an additional cost of £220 million

The world financial crisis of 2007/8 created a 'black hole' in public funding. Finding public money continues to be difficult, if not impossible, for any expansion of light rail, when competing against the needs of health services, schools, police, public space cleansing and other publicly provided services. A cash crisis in California is matched by similar crises in Britain and Ireland.

In these circumstances, it remains important that effective arrangements are in place to hold officers – and consultants employed by them – to account through effective challenge and scrutiny.

These reports have done nothing to bolster public confidence in Britain over the effectiveness of light rail schemes, or that public authorities can achieve value for money. This may go a long way to explain the coolness of the UK Government to fund new schemes, despite their popularity with the public, their ability to divert commuter car traffic and their several environmental benefits.

LJS Lesley
ISBN 978-1-83708-399-2
https://doi.org/10.1108/978-1-83708-398-520261002
Emerald Publishing Limited: All rights reserved

Chapter 2
Characteristics of light rail

This chapter discusses the origin, definition and operation of LRT.

2.1. Abandonment and reinvention

2.1.1 Abandonment

Chapter 1 set out the background to the growth of tramways and light rail systems. Here we consider in detail why in some countries they were abandoned and in others retained and developed, and are now enjoying a period of renaissance. The first tramways or street railways were developed mostly by private enterprise, often linked to property development, enabling urban residents to become more mobile. As the population of the developed world grew more prosperous, mass production of automobiles lowered costs, so people could widen their travel horizons. In travelling by car, people used public transport services less. Attempts were made by private operators in the UK and US to develop and improve their systems in order to compete with private car travel for some urban trips.

In many places tramways were taken into public ownership and, therefore, were constrained by the way public bodies can operate. This included replacing assets as they wore out or became obsolete. Too often, this was seen as drain on public funds, ignoring the operating surpluses previously enjoyed that relieved local taxes. As motor buses matured, these were seen as a cheaper option than tramways and avoided the need to maintain tracks. Local authorities still had to maintain roads, now wearing out faster due to bus services.

In the US another mechanism led to the abandonment of many streetcar systems, namely the National Lines Corporation set up in 1938 and funded by General Motors, Firestone Tires, Standard Oil of California, Phillips Petroleum and Mack Trucks (Black, 2007). This company bought up streetcar systems and converted to bus operations. The main and successful objective was to increase the sale of automobiles, gasoline, tyres and so on. These companies were convicted in 1949 of conspiracy and anti-trust behaviour but fined only a derisory $100 000, a small sum compared to the millions made in the sale of automobiles.

In the absence of street rail systems, and the growth of car traffic, local authorities built new roads to cope with growing congestion, at a cost much higher than retaining tramways. There was no acceptable alternative to car travel. It soon became apparent that this would not solve traffic problems nor was it economic, set out clearly in 'Traffic in Towns' (Buchanan, 1963) and similar reports.

Even today there are critics of light rail in the US, like Wendell Cox (2008), Randal O'Toole and Bill Vincent (of the Breakthrough Technologies Institute), advocating bus rapid transit (BRT) as a cheaper option than light rail. This point of view rests on the erroneous assumption that bus and light rail are perceived as equivalent alternatives by passengers, especially automobile commuters.

Advocates for BRT do so from a 'supply side' perspective, ignoring the demand side, which Chapter 1 considers in depth.

Some transport commentators criticise light rail as being expensive, and that buses (on busways) could do the same job for much less money. This was a policy followed in Houston in the 1980s and 1990s with the construction of over 100 miles (160 km) of busways later modified to high occupancy vehicle (HOV) lanes on many major routes.

More importantly, buses are not perceived to be the same offering by those whose custom is required to earn the revenue needed to pay the operating costs. This is particularly important relating to people who presently drive cars. While there have been ad hoc studies in different cities on the market assessment by residents and, especially, car commuters of their views of alternative forms of transit, none has been as systematic and evidence based as Report 1221 published by the US Transportation Research Board (Tennyson, n.d.).

Report 1221 covers a 40-year period, and studied the impact of closing (light) rail lines on travel patterns, as streetcar and commuter rail lines were abandoned and replaced by buses. In many cases, buses were later abandoned. Towards the end of the period, the impact of opening new light rail lines, to replace buses, on travel patterns was also studied. Report 1221 concluded that the closure of rail lines, replaced by buses, often with enhanced service frequencies, resulted in a loss of transit patronage of between 30 and 40%. The passengers lost to transit transferred to car travel.

2.1.2 Reinvention

Ironically, it was those countries worst damaged during the Second World War (Figure 2.1) that retained and developed their tramways. These included Austria, Benelux, Germany and all the countries occupied by the Soviet Union. Tramways were all in a bad state of damage and lack of maintenance. The countries were economically badly run down and a 'make do and mend' policy was adopted. This saw tramways retained and new lines opened. Partly, trams ran on home-produced electricity, not dependent on expensive or imported oil. As these countries developed their economies, labour shortages became critical, partly due to the loss of manpower during the war and partly as a national priority export industries needed labour.

Figure 2.1 Tramway retained following devastation from Second World War

Workforces saved from running public transport could be used in productive industries. Tramways were ideal for increasing workforce productivity. Three cars forming a tram train needed only four staff, whereas three separate trams (or buses) needed six. The big break through was developing and improving articulated trams (Figure 2.2), having the capacity of three separate cars but needing only two staff. With changes to fare policy, ticket sales were made 'off-vehicle', operating a 'honesty fare policy', backed up by roving inspectors. An articulated tram then only needed one member of staff.

Figure 2.2 Examples of articulated trams in the 1950s and 1960s

Brussels

Gera

Dusseldorf

Budapest

The improvement of vehicles led to the improvement of routes, with the segregation of tracks from other traffic, either by moving tracks off road or by moving other traffic off tracks. When coupled with priority at road junctions, service speeds increased and trams became competitive with car commuting. Examples of different track arrangements were set out in Chapter 1.

The opening of new light rail lines allowed TRB Report 1221 to make a comparison with existing bus lines. Where bus lines were replaced by light rail, on a like for like basis, rail carried about 40% more passengers than bus lines. Rail also had between 34% and 43% passengers who left a car at home or a park and ride station. Report 1221 concluded, 'In most cities served by buses exclusively, transit riding has declined 75% over the past 40 years. Exclusive busways have not made much difference absolutely, but they have helped relatively ... In 11 areas with updated rail transit facilities, ridership has increased markedly, often by more than 100%. In two of these areas, the transit systems are attracting more ridership than they did when gasoline and tires were rationed. It appears that rail transit makes a great difference in ridership attraction, with attendant benefits.'

Report 1221 continued, 'When these service conditions are equal, it is evident that rail transit is likely to attract from 34% to 43% more riders than will equivalent bus services. The data do not provide explanations for this phenomenon, but other studies and reports suggest that the clearly identifiable rail routes; delineated stops that are often protected; more stable, safer, and more comfortable vehicles; freedom from fumes and excessive noise; and more generous vehicle dimensions may all be factors ... Those engaged in the alternatives analyses (*of transit modes*) and similar studies would be well advised to consider these differential factors before making service recommendations or traffic relief assumptions. Future problems with air pollution, congestion, and funding may all be seriously affected by these considerations.' This latter point was prescient when TRB Report 1221 was published.

2.2. Market perception

All enterprises should be aware of the way their products or services are perceived both by customers and those who do not purchase. This differentiation is factored in as part of the marketing process. Repeat purchases are the foundation for a successful product. This is more than just advertising. There needs to be solid market research both to develop and then launch products. Feedback is equally important. Often this comes from dissatisfied customers. Good responses from the supplier can both change perceptions and convert critics into advocates. Marketing needs to be proactive in seeking views of those who use and, importantly, those that do not use the service.

There have been few comprehensive studies like TRB Report 1221 in other countries. Political decisions led to trams (streetcars) being abandoned in Britain during 1948–1962, all replaced by buses. In most cases the new buses were more comfortable than the 50-year-old trams being replaced. Some ran faster with higher frequencies, as the buses carried fewer passengers than the abandoned trams. In all cases, however, the patronage fell immediately by between 30% and 40%, and continued to decline at an average of nearly 2% per year, and ultimately by over 70% since 1955, when bus ridership peaked UK. London now carries about half of all bus patronage in the UK.

Railways in general have a devoted following far outweighing the actual use of rail transport or the market share enjoyed. Light rail and tramways are a specialist part. Much of the present literature on light rail is aimed at enthusiasts. The target market for light rail must be those, a large majority, who do not use urban public transport but drive their own cars. Politicians often think that fares are the main determinant of the use or otherwise of public transport. Given that owning and using a car is more expensive, cost has been found in many market surveys not to be the most important factor, usually fourth or fifth behind convenience, reliability, speed and comfort.

Discussions on the perception of tramways have been set out in a large number of publications. Those aimed at enthusiasts should not be dismissed. Many enthusiasts have encyclopaedic knowledge of light rail matters and often useful insights into how light rail might be improved (e.g. Buckley, 1975; Cudahy, 1995; Dunbar, 1967; Garbutt, 1989; Hass-Klau, 2004; Joyce, 1964; Powell, 1997). Identifying and applying this valuable information is however a different matter.

There is also much material on the World Wide Web (e.g. APTA, Lightrail, Lightrailnow, LRTA, Transport 2000, UITP and Wikipedia). Much of this is ad hoc and difficult for a practitioner to apply with any confidence. This is important when trying to justify, promote, plan, design, implement or operate a new light rail system. This present book is aimed at those charged with designing, building and operating light rail systems, and is based on considerable practical experience, research and discussions with other knowledgeable practitioners.

Many transport professionals often consider these matters from a supply side perspective (Coffey, 1992; Lesley, 1991). Car ownership was low in North America in the 1920s and Europe in the 1950s. Such an approach then made sense, since tramways, streetcars and light rail systems were selling to 'captive' riders. This is particularly true of those texts that try to show why one mode of public transport or transit should be used in one circumstance but not in another (Tolley and Turton, 1995) (see Table 2.1). This is usually based on peak traffic forecasts. In particular the debate about bus, light rail and metro falls into this category, and is based on the fallacious assumption that potential passengers consider all these modes to be of equal merit and quality. Market research shows this clearly not to be the case (Lesley, 1974).

A supply-side perspective is, however, not appropriate in developed countries with high levels of car ownership (Table 2.2) where public transport is a minority carrier of personal trips, even in major cities like Berlin, London, New York, Paris, Toronto and Washington (Knupfer *et al.*, 2018). For the promoter of transit improvements to approach the subject with a supply-side perspective risks missing the target market and ending with low ridership.

The US Transportation Research Board Report No. 1221 (Tennyson) offers evidence for the promoters of light rail, about the level of patronage that can be attracted. This was the result of analysing 40 years of transit data in the US. On the other hand, the UK National Audit Office Report on Light Rail (NAO, 2004) should make salutary reading for all those promoting light rail, from the pitfalls of over optimistic patronage forecasts and cost over runs.

This present book tries to redresses that imbalance by looking at the considerable market research on what is needed to get car commuters onto light rail transit through attraction, rather than punitive measures like parking restrictions and pricing, congestion charging or other methods of real-time road-user taxes. It also considers how to deliver light rail systems that can generate enough

Table 2.1 Capacity of urban transport modes

Mode	Max. Capacity People/hr/lane	Av. Speed (km/hr)	Stop Distance (km)
Bus on road	9000	16	0.2
Bus /busway	20000	56	0.8
LRT	40000	26	0.5
Metro	60000	32	1.5
Car on city road	1000	19	–
Car/freeway	3000	72	–

From Tolley and Turton (1995)

Table 2.2 Levels of car ownership in developed countries (2022)

Country	% households with car	Cars per 1000 population
United States	92	908
Canada	72	790
Australia	91	782
Italy	89	755
Poland	65	687
France	86	668
Japan	81	661
Germany	77	628
Spain	70	627
UK	68	600
Belgium	73	590
Netherlands	74	588

operating surplus and, therefore, not need operating subsidy from taxes. As importantly, the farebox should generate sufficient surplus to depreciate assets for replacement when due, on a financially sustainable basis.

France offers evidence from the opening of some 30 new light rail systems since 1985. Nearly all replaced the busiest city bus lines. On opening the new light rail lines, the patronage increased within a year by a minimum of 30% when compared to previous bus services. Remaining bus services were reorganised to provide feeders for light rail lines. Bus services also enjoyed an increase in ridership when part of a through journey is by tram. Their success was helped by the inclusion of purpose-built interchange stops and timed connections.

In Britain six new light rail lines have opened since 1992. Most of these have used old or closed urban rail lines. Market research of the patronage of these lines has shown that between 20% and 40% of passengers have transferred from car commuting, with a UK average of 25%.

All these data indicate that there are qualitative differences between light rail and bus services, which heavily influence customers, and therefore ridership and revenue. In particular, light rail has shown itself able to attract significant traffic from private car commuting as demonstrated in San Diego (Figure 2.20). This modal transfer is critical to help cities address and reduce traffic congestion, air pollution and other environmental problems. In Houston, three more LRT lines (Figure 2.23) have opened to supplement the successful first line in 2004. Ridership on these is now 18 million per year. Two further lines are planned with rising public support.

2.3. Tramways incremental development into light rail

The horse and cart have long been replaced by pickup trucks. Similarly, streetcars running on street mixed with other and congested traffic should now part of history (Figures 2.3–2.6). This is discussed more fully in Chapter 1 (see Sections 1.2 and 1.5).

Figure 2.3 John Stephenson – pioneer New York street railway 1832

Figure 2.4 1887 First electric railway, Berlin Industrial Fair

Figure 2.5 Trailing wire pick up (Troller) Orange, New Jersey 1887

Photo C.S. Dunbar

Figure 2.6 Sprague's trolley pole system, Richmond Virginia 1888

Photo C.S. Dunbar

The application of electricity was a step change in the development of street railways, since this reduced costs and increased speed, and rescued operations from the vagaries of the health and availability of horses.

Transmitting electricity to moving trams was quickly solved. First came the troller system (Figure 2.5) used in Orange, New Jersey, in 1887, but its junctions were difficult. Sprague solved this with trolley poles contacting the underside of a live overhead cable, first demonstrated in 1888 in Richmond, Virginia (Figure 2.6), and still serves the role of transferring electric current to moving vehicles, for a few systems – for example, Boston and Toronto, some industrial cranes and all trolleybus/coach systems. Later, bow collectors, developed in Budapest (Hennessey, 2008), were adopted by a number of systems including Aberdeen, Dundee, Glasgow and Leeds in the UK, as well as Ritten in Italy and Plauen in Germany. At most terminals, bow collectors (Figure 2.7) automatically reverse for a return trip.

Later tramways adopted pantographs (Steimel, 2007), originally developed for railway operation. Pantographs (Figure 2.8) operate equally well in both directions. Both bow and pantographs meant that the overhead lines could be simplified, and dewirement became rare.

The use of electric traction significantly reduced operating costs, and made streetcars and tramways the first mass means of urban transport affordable to even to those with low incomes. This created a new market of urban residents who previously had to walk.

In much of Europe and North America, the funding for the new electric streetcar and tramway systems came primarily from private investors, with commercial companies seeking to reward the capital from operating revenue. In some places this led to accusations of profiteering and the intervention of public authorities, firstly to regulate fares and then to take systems into public ownership when cash strapped operators could no longer maintain equipment.

Figure 2.7 Bow Collector on Leeds Tram prior to 1959

Figure 2.8 Pantograph, City Class in Birkenhead

This changed light rail in most places from a commercial service, to systems having to satisfy political aspirations, especially at election times. Fare rises were seen to be a vote loser. Public bodies often used the operating profits from these systems to subsidise municipal spending, while at the same time rarely was enough surplus retained for depreciation and asset replacement – for example, in Liverpool during the 1920s and 30s, when on average of some £200 000 per year was taken from the tramway account to reduce local property taxes, effectively a subsidy for those with large properties. And, during this time, the Chief Engineer built a new fleet of trams, claiming they were refurbishments, so could be funded from revenue, rather than capital accounts. Only the fleet numbers of the old trams were retained.

Similarly in Europe, tramways were often used for social policy by, for example, making slum clearance possible, allowing people to be moved from inner areas to new housing areas in the suburbs. Municipal trams allowed the people moved from slum areas to get to their work in the city centre, often at subsidised fares.

By the mid 1920s automobiles were serious competition for streetcar lines in the US, still operating vehicles nearly 30 years old (Figure 2.9) and antiquated by comparison with contemporary cars. The private streetcar systems co-operated to build a modern vehicle that could out compete automobiles, not just in comfort but also in speed. The outcome of this programme was the radical Presidents' Conference Committee (PCC) streetcar, which funded the Transit Research Corporation (TRC). This had smooth acceleration and braking, a high speed, good heating and ventilation, comfortable seats and looked as good as the then modern automobiles.

The PCC design was licenced by the TRC to car builders in the US and some 5000 were built after 1930 (Figure 2.10), which modernised many of the remaining networks in North America, extending the service lives of many systems. The author can remember visiting Boston in 1966 and being allowed to drive a PCC on the Green Line, which was used for staff training. This was one of a pair in a 'consist' (train) in a multiple unit configuration. The emergency brake was formidable and very effective. After the Second World War the TRC licenced a number of European car builders to build the PCC and derivative vehicles. Until 1990 some 60 000 further variants of the PCC were put into service on many systems.

In North America many streetcar systems remained in private ownership much longer than in Europe (e.g. Figure 2.9), and indeed in the 1940s some were bought out by the National City Lines, which was funded by oil, tire and automobile interests. These streetcar systems were then run down, closed and replaced by buses. This anti-trust behaviour was investigated by the Federal Commerce Commission. Years later it imposed a derisory $100 000 fine, by which time transit use was in sharp decline. Ironically, the streetcar systems that survived – for example in Baltimore,

Figure 2.9 Private operator in Philadelphia 1935

Source: R. Hill collection, courtesy of Ken Josephson

Figure 2.10 Examples of PCC streetcars in North America

Toronto Pittsburgh San Fransisco Chicago

Boston, Philadelphia, Pittsburgh, San Fransisco and Toronto – did so due to the innate conservatism of their municipal owners, although in all cases the systems were mere shadows of their former networks.

When worn out infrastructure or rolling stock came up for renewal, some public operators, particularly in Britain, France and North America, replaced these with cheaper and more 'flexible' buses. Sometimes the hybrid trolleybus (coach) was used but more often the choice led to electric traction being abandoned. In the UK and some other places tramways were often powered by municipal power stations, supplying power on a marginal cost basis. When these were nationalised in the UK in 1948, the cost of electricity increased and reduced the financial viability of municipal tramways.

When oil was cheap and little was known about the health impacts of air pollution or the limit of oil reserves, this seemed a sensible policy, if it were not for one factor. It ignored the wishes of tram passengers and the aspirations of the market. Added to this was that many politicians drove automobiles and often saw streetcars as holding them up and causing congestion. This political prejudice cannot be ignored in the history of the decline of the first generation of street railways.

The closure of tramway and streetcar systems led to a reduction of patronage, typically about 30%, with many ex-passengers buying and using motor cars. This created traffic congestion and pressured highway authorities to build new roads and enlarge existing streets, at costs orders of magnitude greater than renewing worn out rails or buying new electric rail vehicles.

Even in a command-economy city like Budapest in the 1980s the performance of slower on street tramways delivered quicker door to door journeys than faster metros (subways) (Figure 2.11), because of more accessible stations and the short waiting times for frequent services. As a senior transport professional observed, trams give 90% of the benefits of a metro at 10% of the cost. In Budapest with a much larger tram network than three metro lines, the network also offered faster door to door travel.

Arguments for such systems to adapt to high (Western European) levels of car ownership led to the adoption in Budapest, before the fall of the Berlin Wall in 1989, of a Gothenburg style zoned city centre, to reduce (through) car traffic, and measures to protect tram tracks from car traffic to maintain operating speeds (Lesley, 1986).

By the 1970s, North American cities were realising the truth of Professor Sir Colin Buchanan's seminal report 'Traffic in Towns' (Buchanan, 1963), advising the UK Government on future policy for urban transport. He showed that the latent demand for private car trips was so high, even in low density cities like Los Angeles, there could never be enough road or parking capacity to meet peak travel flows entirely by car. He advocated a twin approach of better public transport and steps to manage car traffic to levels that could be environmentally and socially accommodated. With North American cities reaching gridlock and down town areas dying for want of acceptable access that bus systems did not offer, municipal authorities saw that although rail transit attracted car commuters, new subways or metros were unaffordable.

Metro or subway systems are expensive. The reasons for this relate to the need to acquire large sites for stations and the cost of tunnelling. Pierre Laconte, then Secretary General of UITP, observed

Figure 2.11 Journey time comparisons in Budapest 1985

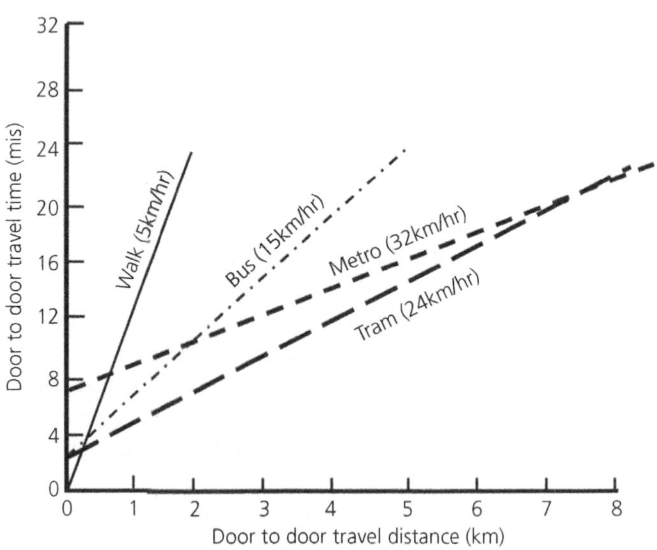

Source: The author

that light rail offers 90% of the benefits of a subway at 10% of the capital cost (Laconte, 1978). Light rail or modern streetcars are quick to install and an affordable solution. The image of streetcars in North America bore antiquated connotations with ideas of slow and inefficient operation. So, a new generic name 'light rail' relaunched street railway transit.

By the 1980s policies of privatisation and deregulation were implemented in many countries. The aim was to remove the patina of public sector inefficiency and pork barrel politics. We will not debate the rights or wrongs of either public or private sector as the promoter or provider of light rail systems but show how light rail can be made to work through either approach.

The use of open competition is now a cornerstone of the procurement policy for publicly funded projects, including those to improve or reduce the cost of transit services all over the world. In North America transit remains highly regulated and largely publicly owned. There is also a debate over the difficulty of introducing innovation in light rail when risk-averse public authorities only seek equipment that is tried and tested. Had this been the case 150 years ago, would horses still be pulling streetcars?

Nevertheless, there has been incremental improvements in vehicles, see Figure 2.1, and track alignments (see Chapter 1 section 1.7). These have improved the productivity of services and reduced costs. Faster speeds attract trips that would otherwise be made by car. One of the fundamental changes has been to adopt coarser networks operating at high frequency, with short waiting times. Using purpose-designed, passenger-convenient interchanges between lines enables many journeys to be made quickly, without the need to park, except perhaps at an out-of-town park and ride stop. A US transit 'rule of thumb' states the waiting time should be less than the in-vehicle time. Even a coarse 'few to few' network can have stops within acceptable walking distance. With park and ride much larger catchment areas can be served.

Overhead line (OHL) power supply, using a variety of devices, latterly derivatives of pantographs, is the most widely used system. There are also proprietary ground level and underground systems. Vehicles powered only by batteries, and more exotic sources like hydrogen and bio-methane are also being developed, but require recharging during the day for some systems. Unless there is an unexpected technical breakthrough, OHL remains the most economic and reliable power supply system. There is the added advantage of OHL in being able to supply high currents for hill climbing and acceleration, which is not possible in vehicles with on board power systems.

With the right in-vehicle traction control, braking through regeneration can recover some of the acceleration power for nearby vehicles. The claim that OHL is visually unacceptable can be disproved in those cities which have extensive networks. There is also the question of trade off. In the 1960s a campaign was started in San Francisco to abandon electric street transit with OHL in favour of buses, to remove 'unsightly' OHL. The city council held a referendum which was based on 'OHL against diesel fumes'. Over 70% of electors opted to keep the OHL, a wise decision now that the health impacts of motor vehicle toxic emissions are better understood.

2.4. Alignment and locations

In designing a new tram line, and especially a new system, there are a number of factors to consider. The strategy for installing a new light rail system should be to build the most financially viable line first. Additional lines can then be added on a marginal cost basis, as a depot for maintenance and repair can be shared, as can the central management costs. This means that less viable lines can expand the network. Then network synergy increases the traffic, by linking the existing line with new, improving the performance of the whole network.

In getting new light rail lines open, the key is to minimise the capital cost (CAPEX) by efficient construction, equipping and commissioning. When light rail schemes are being promoted by public authorities, the capital required has to come from a budget provided by taxpayers. This is not limitless. Demands on taxes for public capital works and revenue payments, like transit subsidies, usually outstrip taxpayers' willingness to fund. Inevitably a mechanism to justify the use of 'free' money and maybe also subsidising fares is needed (Lesley, 1987). Benefit-cost analysis (BCA) was developed as one way to demonstrate and justify the non-financial benefits that such public investments confer.

The benefits in BCA include environmental improvements, reducing congestion, encouraging economic activity and reducing the need to expand the road network. These, however, produce no income to offset the transit investment or operational costs. Politicians sometimes undertake an electoral calculus about which projects will win the most votes. In this scenario light rail has a low priority compared to schools, hospitals and social care. Even in cities with widespread traffic congestion or high levels of toxic traffic air pollution, there has not been enough pressure to provide light rail for more than a token diversion of car trips. Highway capacity is rarely reduced in such cases. This means that suppressed car trips can quickly fill the space vacated by trips attracted to light rail. No overall environmental improvement can be measured. There are texts by transport economists that can enlighten this debate (e.g. Glaister, 1981; Lesley, 1996; Townsend, 1969).

BCA has however been systematically criticised, by among others the 'Strong Towns' movement, on the basis that such public expenditure rarely increases the city's (tax) income by enough to retire the transit (CAPEX) debt. Indeed, the city's finances may have been worsened to such an extent that other vital services, like repairing roads, are neglected.

New light rail line promoters must therefore ensure that at the very least the patronage and revenue exceed the operating costs, so there is no need for an operating subsidy.

2.4.1 On street track and under street utilities

Streets are historic and natural channels for movement. Streets with the highest level of movement must, therefore, be prime candidates for a new tramway or light rail system. This can be backed up by analyses like origin-destination trip distribution, to determine the total volume of potential traffic for a new rail system in a particular corridor. The question then becomes balancing the needs of existing street users, both pedestrians and vehicles, with the under-street utilities that service properties alongside the route and may require relocation.

By definition a new rail line will be a major carrier of passengers. Various graphics demonstrate that a light rail vehicle (LRV) (tram) occupies much less road space than the same number of passengers in private cars (Figure 2.12). Roads also provide access to adjacent buildings for the delivery or collection of goods. The siting of a new tramway needs enough space to be provided for the servicing of buildings alongside the road. Parking is, however, not an efficient use of road space, certainly not on main roads, so should be discouraged, and so keep all traffic moving.

Figure 2.12 Tram passenger capacity compared

These 170 cars carry 200 people, who could all be in one LRV
Source: Bath trams website

Access to the new tramways is achieved by providing tram stops. These should be designed especially for passengers with mobility impairments, which can be particularly challenging. Stops will need a platform at least 2 m (6 ft 6 in) wide, the length of the operating vehicles and the same height as car floor entries, to provide level boarding and alighting. In some places this may best be accommodated by the use of an island platform in the middle of the street, with dedicated pedestrian crossing places at either end (Figure 2.13).

This means that waiting passengers will not obstruct people walking on the footway (sidewalk/pavement), To achieve this the rail tracks will deviate from a central road alignment, and may also require the tracks to be adjacent to the road kerbs. Island platforms can also be installed in narrow roads, where footway space has important other uses. Figure 2.14 shows such an example.

Figure 2.13 Island platform schematic

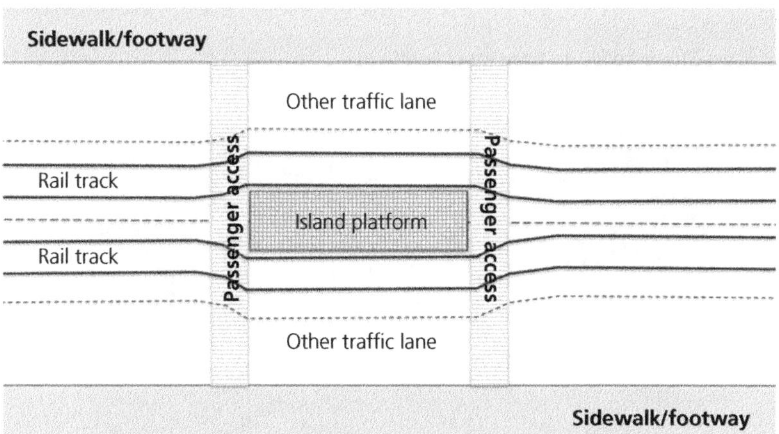

Source: The author

Figure 2.14 Island platform in narrow road – Edinburgh

Source: Google Streetview

In narrow roads, perhaps with only two traffic lanes, the stop platforms will have to be part of the footway and, to make rail operations reliable, most of the through vehicular traffic will be diverted to alternative routes. There are many practical examples of this (e.g. Figure 2.15, which shows the rail tracks occupying the two traffic lanes, shared with other traffic, including buses, and the West Street tram stop on the roadside in Sheffield, UK).

Next comes the thorny issue of the under-street utilities (Figure 2.16). There was a landmark case in West Ham (UK) in 1903 when the council started to electrify the inherited horse tram tracks. The local gas and water companies wanted their pipes moved away from the tram tracks, and for the council to pay, which it refused. The case went to arbitration, which found in favour of the council (Collins, 2001). Since then, the under-street space has been filled with even more utilities, including foul water drainage, electricity distribution and telecommunication cables, which now enjoy 'grandfather' rights of reasonable access to plant for maintenance and repair.

Figure 2.15 West Street kerbside tram stop Sheffield

Source: Google Streetview

Figure 2.16 Example of congested under street space

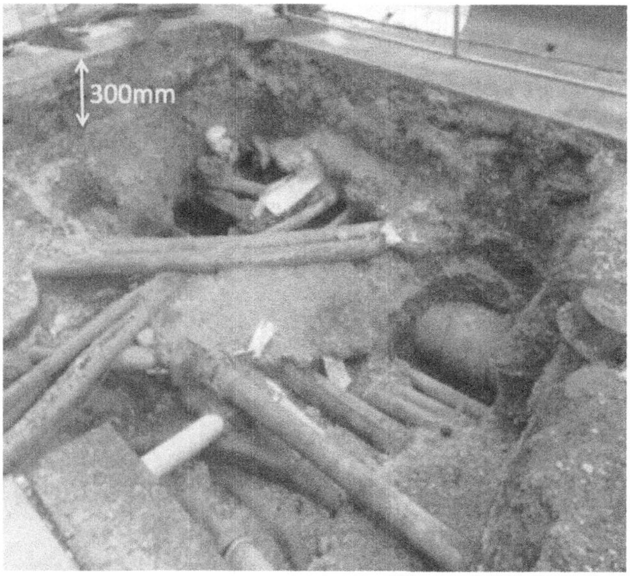

Source: The author.

Note that the pavement thickness is 300 mm (12 in) and the most shallow plant (red ducts) is another 300 mm (12 in) deeper.

Most street track forms have rails sat on a concrete raft about 6 m (20 ft) wide and 500 mm (20 in) thick (Lesley, 2025). Any utility plant under this raft will be inaccessible, without the considerable expense and difficulty of drilling through the concrete and disrupting the rail service. In the UK, therefore, light rail promoters have been required to relocate under street utility plant away from the track path. This can be over half of the budget for track installation, and takes a considerable time using specialist contractors. By and large, the utility companies prefer to leave plant in situ, provided access is maintained for repair and maintenance, since relocating always creates secondary and inter-face problems.

A considerable effort has been given to developing new street track forms, which claim to obviate the need to relocate utilities. This is discussed at some length in 'Embedded rail tracks' (Lesley, 2025). The author's own invention of 1989 (Figure 2.17) avoids utility relocation and has been accepted by most utility providers. It has also been in continuous use since March 1996 in the tramway in Sheffield, where it replaced girder rail track that failed in less than six months service, from the impact of heavy road vehicles. In that time, the single-track installation with trams every three minutes has shown minimal wear, and has not needed any maintenance. The bottom of the rail groove has compacted mud 20 mm (0.8 in) thick. At that rate of wear, the LR55 installation looks good for about another 20 years.

Figure 2.18 shows that LR55 needs a minimum of road excavation (*c.* 5%) compared with track that needs to relocate utility plant and install a foundation slab. Protocols for access to such plant

Figure 2.17 LR55 track, compared to 60Ri rails and Sheffield installation

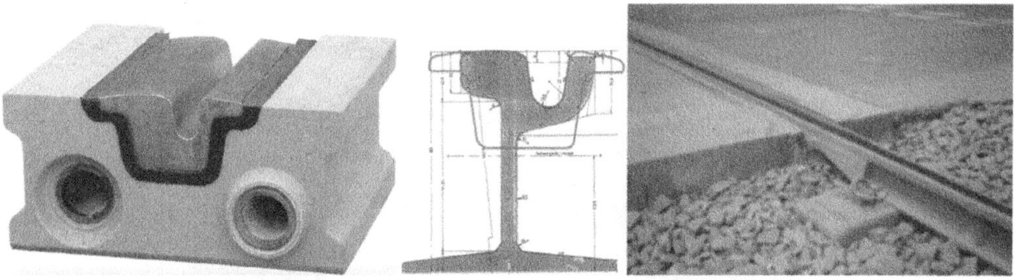

Figure 2.18 Typical LR55 section in an existing road pavement

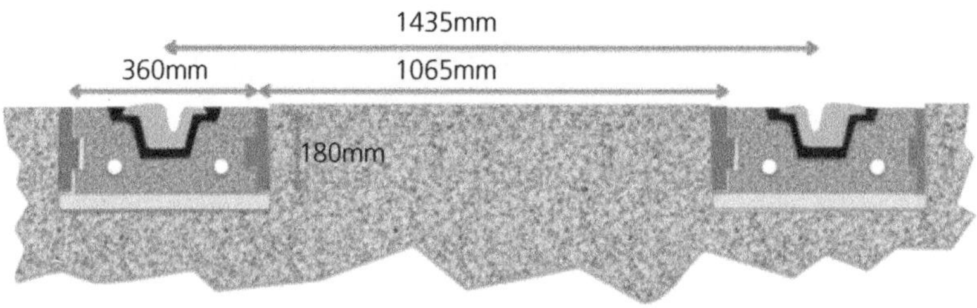

under LR55 track, while maintaining a rail service, have been developed. LR55 track can also self-support over trenches up to 1 m (3 ft 4 in) wide, making utility maintenance faster and less expensive.

2.4.2 Off street

Installing light rail tracks off street presents a different set of challenges (Figure 2.19). Often there is no drainage, except naturally through the ground. Additional drainage may need to be provided to avoid tracks becoming waterlogged or flooded in wet weather. The ground may not be robust enough to bear the static load of the track and the dynamic load of passing trams. The procedure for laying conventional railway tracks is well known and the components readily available. Often conventional railway tracks are used for off street light rail lines, with plain rails on tangent/straight tracks, and grooved rails on small radius curves.

Using current railway practice (a) means that the bottom of the sleeper/tie only has about 5% of its base supported by ballast. Dynamic vehicle loads create high levels of stress and the ballast lumps begin to disintegrate into powder. This clogs the rest of the ballast, which loses its drainage capacity. On railways this leads to the need for periodic cleaning or replacement of the ballast.

With a highway approach (b) laying an asphaltic layer on the graded ground (base) provides a temporary roadway for materials to be brought to site. When the sleepers/ties are laid, their bases are nearly 100% supported so there is little dynamic degradation.

Figure 2.19 Possible off street light rail track

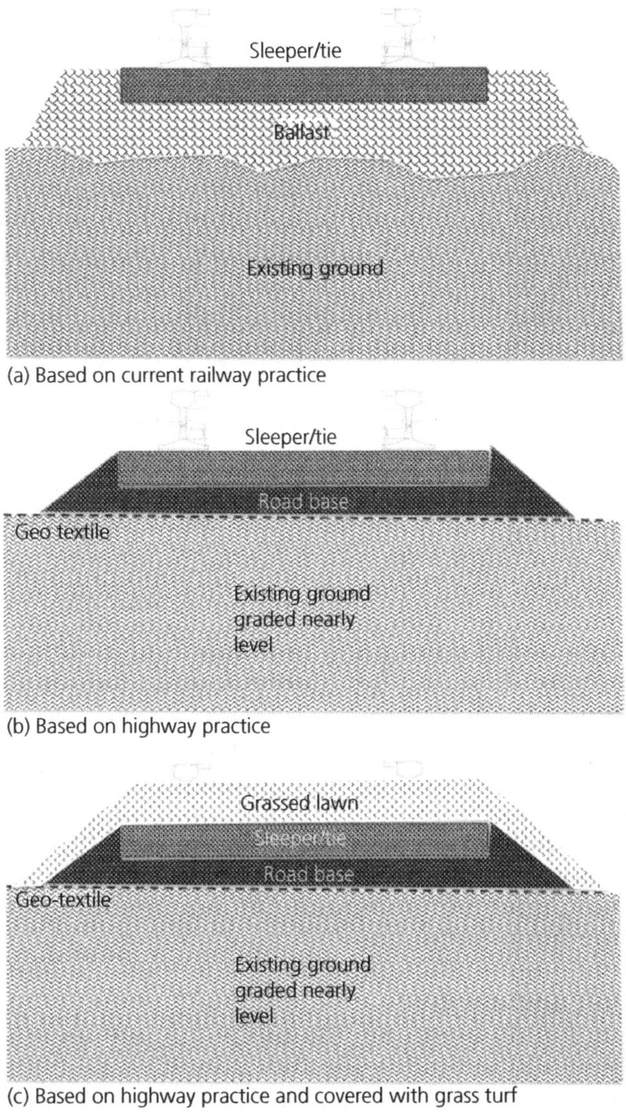

Option (c) takes (b) and softens it by laying grass turf to near the rail head (Figure 2.20). Grass tracks were pioneered on the Liverpool tramway in the 1920s.

Over time wild flowers will blossom, and passing rail vehicles keep the vegetation to a reasonable height.

Using grass in this way has a number of other benefits. The grass helps to remove toxic gases from the local polluted air. The mass of the turf dampens the rail webs and reduces airborne noise by up

Figure 2.20 Example of off street light rail track, before and after grass

Source: Institute for Agricultural and Urban Projects at the Humboldt-University, Berlin 2010

to 4 dB(A) compared to ballasted tracks. This is a significant noise reduction. The turf absorbs rain and releases it slowly, reducing ponding or even flooding. Grassed tracks have lower maintenance costs and people like to see greenery, which has a calming effect. As well as encouraging wild flora, the grass also provides a home for wild fauna, enhancing the local ecology and habitat.

As well as possible environmental advantages, if the off-street line provides a direct route between major generators and attractors of traffic, a faster speed is possible, which will maximise the patronage for the light rail service. Off street routes, whichever track system is adopted, are less costly to maintain than on street tracks, which are impacted by heavy road vehicles.

2.4.3 Power supply

The majority of tramway and light rail systems use an overhead (OHL) power supply cable with a current collector like a pantograph on the roof of vehicles. Historically, other power supply systems have been used, including ground and underground. OHL has proved to be robust and economic. This is considered more fully in Section 4.8.

2.5. Operation and control

There are two levels to consider for operation, the strategic and daily service provision. The strategic level includes identifying market share and likely levels of patronage. From this, timetables, vehicle and staff rostering can be drawn up. These will consider the vehicle maintenance rotation and ensure that the rolling stock is uniformly used and worn, so that replacements can be planned well in advance. The design and specification of these will reflect the experience of operations. This can be incorporated together with a reflection of passenger feedback in placing an order for new stock.

The daily service is what passengers see, and includes valeting the vehicles inside and out before the start of service. Most tramway and light rail systems operate on a 'line of sight' basis, with drivers fully responsible for safe running. New drivers will have been trained in 'defensive techniques', since tram drivers must anticipate the movement of other road vehicles and pedestrians. The vehicle speed is the only safety control available and drivers must be prepared to stop at any time, both in anticipation of a potential impact and in emergencies.

All rail vehicles running on roads will be fitted with emergency brakes, usually magnetic track brakes, to enable trams to decelerate at the same rate as other road vehicles. Typically, this is about 0.3g. Part of driver training will be in assessing what the impact of emergency braking will have on standing passengers. When emergency braking, many passengers will fall over even if holding onto a pole. Injuring a large number of passengers needs to be avoided, by anticipating and service braking early at about 0.15g, which even elderly standing passengers can accommodate.

In driving on line of sight in streets, tram drivers must also obey traffic regulations, including traffic signals, if these are not pre-empted to green to allow junction crossing without stopping. This is a common priority on many systems, pioneered in Zurich in the 1960s. When the tramway turns off the road, the tram driver will need to signal this to other road users, since, and especially at track junctions, it may not be clear to other road users which way the tram will go, even with the tracks in the road. This can be mitigated by the use of dedicated traffic signals for trams and clear road markings.

Trams will also be fitted with an audible warning, to ensure that pedestrians, particularly, are aware of the approach of a vehicle much quieter than other traffic. Some systems have two different audible warning devices, one quieter to announce presence, the other loud to warn of possible danger. The latter is used on segregated tracks, where faster speeds means a longer stopping distance.

To operate a reliable service, the vehicle running timetable should realistically reflect the running conditions and time needed to pick up and set down passengers. This will include the usual traffic conditions experienced, and if there is no priority over other traffic, the speed of general traffic movement.

There are many examples where the right combination of service speed and frequency has ensured that new light rail services gain a large travel market share in the corridor served – for example, San Diego (Figure 2.21) in the US, where the first line was opened in 1981, replacing a bus route.

Figure 2.21 Light rail replaced bus service in San Diego

Photo Peter Ehrlich

Figure 2.22 Impact of first San Diego light rail line

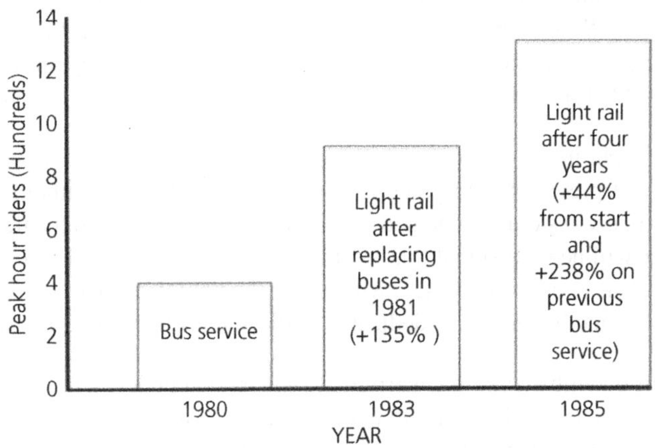

Source: TRB Report No. 1221

Figure 2.22 shows that, nearly two years after opening, this light rail line had increased peak hour patronage from 400 by bus to nearly 900, and after a further two years this had further increased to nearly 1400. As these are commuter trips there can be no question of bus passengers making additional journeys. The increase came from attracting trips that had previously been made by car, and trips not possible prior to the provision of transit service. This is a major conclusion of TRB Report 1221.

As well as growing the share of traffic carried by public transport, light rail vehicles were popular with both passengers and the general public. By reducing the number of buses and cars, a measurable improvement in the environment along the route occurred.

As well as the physical operation of rail vehicles in streets, there is the wider consideration of how to control a whole system. With GPS tracking of vehicles, it is now possible at the control centre to monitor in real time the actual location of all vehicles on the network. The system controller, usually in the central depot, also monitors and controls the power supply and is in touch with the emergency services.

Many light rail systems not only have CCTV coverage at stops and at key traffic intersections. Vehicles can also have CCTV coverage, both internal and external. The controller can therefore be aware of and take measures to prevent disruptions of services. This will be assisted by vigilant drivers keeping control informed by voice radio of possible problems ahead. Should there be an incident inside the vehicle, this can be dealt with at the next stop, where required emergency services can be waiting.

This level of co-operation also needs periodic training and exercises both for the rail operator to implement suitable measures and for the emergency services to speedily and safely resolve problems. The most usual of these are motor vehicle impacts, requiring drivers and passengers to the rescued. A rail vehicle being heavy and more robustly built than a motor car, the smaller vehicle is the one most likely to be badly damaged. Far less common are collisions with heavy road vehicles which can result in the rail vehicle being derailed.

2.6. Equipment and standards

There are two basic sets of standards. One relates to the track and other infrastructure, the other to the vehicles and the modes of operation of the service. Some of these are unique to particular countries or cities, others are more universal.

2.6.1 Track and infrastructure

Starting with the track, most tramways and light rail systems are built to a track gauge of 1435 mm (4 ft 8.5 in). There are however other track gauges in use for historic reasons. Many European tramways operate on 1000 mm (3 ft 4 in) track gauge. Similarly, there are 1524 mm (5 ft) gauge systems. Baltimore and Philadelphia operate on 1588 mm (5 ft 2.5 in) and Toronto on 1495 mm (4 ft 10.88 in).

New systems will almost certainly be built to the 1435 mm standard track gauge, since equipment is readily available. There has been a fallacy that a narrow-gauge track occupies less road space. This is only true if narrow (e.g. 2200 mm/87 in wide) trams are used. The majority of trams being built now are 2650 mm (8 ft 8 in) wide. The logic for this is that most large road vehicles are built 2500 mm (8 ft 2.5 in) wide, and being manually steered occupy an effective road width of at least 3600 mm (12 ft). A rail guided tram 2650 mm wide will safely operate in a lane width of only 3000 mm (10 ft). So can make better use of the existing road space, while providing a more comfortable vehicle for passengers?

The maximum axle load of most trams is about 10 t (22 000 lbs), a little less than heavy road vehicles at about 10.5 t (23 100 lbs). Therefore, rail tracks in streets need to be able to support both rail and road vehicle static and dynamic loadings. In order to achieve that robustly, the foundation of the track needs to be able to transmit these surface forces and distribute them at lower stress levels into the road base, which may be the original ground. The California Bearing Ratio (CBR) test will confirm if the existing road structure is capable of supporting such loadings. Normally a visual inspection of a road will suffice, since if there is a bus service, or the road is used by heavy goods vehicles, and there are no signs of failure – from rutting or subsidence – then the road structure will support a rail track. The use of a CBR test is also important for tracks constructed off street.

Can new tramway tracks support the existing flow of heavy road vehicles, including buses without failure? This is addressed more fully in the author's earlier book on 'Embedded rail tracks' (Lesley, 2025). In summary, there are designs of tracks that are intrinsically able to support heavy road vehicles without failure. These often have long maintenance-free lives under heavy road vehicles without failure. This can be determined by calculation, and not based on opinion.

Added to this is the question of under street utility plant – for example, cables, ducts, gas and water pipes, drains and so on (Figure 2.14) – which from time to time need repair or maintenance. This was discussed in Section 2.4.1. Most track designs have rails resting on concrete foundation slabs that can be at least 3 m (10 ft) wide and up to 500 mm (20 in) thick. Effectively this sterilises the under-street space, since access to any plant underneath the slab will require breaking through the slab and disruption of the rail service.

This is a reason why new tramways often relocate the utilities away from the tracks, although most utility providers prefer to leave plant in place, if access can be provided. Just to install a foundation track slab, a road at least 7 m (23 ft) wide and 700 mm (28 in) deep must first be excavated. This often needs hand excavation, since utility plant can be fragile or dangerous. If utilities are also relocated then the depth will increase to 2.5 m (98 in) deep, creating a logistical problem of what to do with the spoil, which in many countries will be designated as hazardous.

Reinstatement of the road carriageway then requires a proper degree of compaction to avoid subsidence, or the creation of voids under the track slab.

Providing platforms at stops for low floor (300–350 mm/12–14 in top of rail (TOR)) vehicles does not present such a difficult engineering design or installation challenge, since the dead weight of the platform, typically 300 kg/m^2 (425 lb/yd^2), will be much less than the loading of vehicles on rails. Even allowing for a canopy, whether supported by pillars set into the road pavement or integral with the platform design, will not significantly add to the stress into the road pavement.

Finally, for those systems using overhead line (OHL), electrification, supports for the wires and the supply cables must be provided. There are two main systems. One is free-standing poles, either at the side of the track or central between the tracks. Such poles, historically, have been sunk into the ground to a depth of about 1800 mm (6 ft) with a rake on side poles of about 150 mm (6 in) away from the track to compensate for the pull from span wires and the power supply cable. More recently, the railway practice of large concrete 'gravity' blocks set into the street has been adopted by some systems, with the poles having a flange at the base bolted to the block.

Most poles are steel, although there are examples of wooden poles more than 50 years old in city streets around the world. The advantage of wood is that it is more environmentally friendly than steel (Figure 2.23) in capturing carbon dioxide (CO_2) from the air. Thus, a 5 km (3 mile) route with steel poles is likely to add about 43 t of CO_2 to the atmosphere, whereas wood will remove 30 t.

Some systems have used lattice steel masts, while others employ concrete posts. Much will depend on the costs involved and the acceptance by the local community.

Figure 2.23 Comparison of the carbon impact of steel and wood

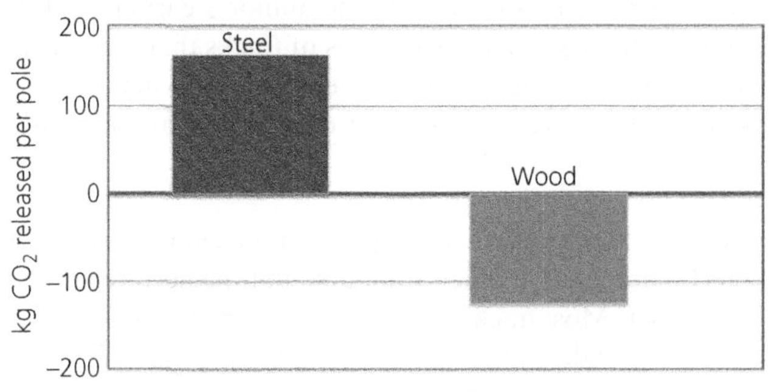

OHL can also be supported from span wires, directly fixed to suitable buildings alongside the route (Figure 2.24). Given the spacing of span wires, about 30 m (100 ft), the OHL designer will have a choice of suitable buildings from which to anchor the wire. A typical pull out force is about 1200 N (300 lb) force. Normal masonry walls can easily support such forces and, compared to earlier elaborate metal plates, simple stainless steel eye bolts bonded into the wall using epoxy or similar high strength cements are more than adequate. Obtaining building owners' agreement can be problematic. The OHL however does not need to be fixed to every building – so there is flexibility. If one owner does not agree, the next might, especially if the alternative is a traction pole outside the building.

To supply the OHL, electrical substations are needed that transform high voltage supplies to the lower voltage and rectify to direct current used for tramways and light rail systems. The substations

Figure 2.24 Span wire wall attachments (old and contemporary)

Croydon (1903) Sheffield

Nottingham

will also need switches to enable sections to be isolated in emergencies or for repair works to be undertaken. These can now be operated automatically and remotely from the control centre.

2.6.2 Vehicles

Tramcars or light rail vehicles need to be road worthy, both in terms of keeping up with other traffic and also being able to survive crashes while protecting passengers. The equipment needed to power and operate such vehicles is well developed. These consist of motors, controllers, current collectors, switches and auxiliary equipment – for example, saloon lighting, brake actuators and a low power emergency supply in case of a failure of the main power supply, enabling vehicles to stop and at night be illuminated so passengers can safely alight. The equipment will also have protection to prevent passengers and staff receiving electrical shocks. These measures are well understood, as are the protection standards required.

Considerable ergonomic research has been undertaken on the safety and the comfort of passengers in moving public transport vehicles on streets. This includes the design and spacing of seats, the number and position of hand holds, both to allow passengers to progress inside vehicles, as well as for the safety of standing passengers at peak periods. This ergonomic research also covers the acceleration and rate of change of acceleration (jerk) that standing passengers can safely accommodate. This work indicates that a maximum acceleration rate of 1.8 m/s^2 (6 ft/s^2) is acceptable and safe.

Similarly, a maximum safe jerk rate of 1 m/s^3 (3 ft/s^3) has been found to be safe. Many contemporary light rail vehicles have acceleration rates much lower than the maximum quoted above. The same analysis also applies to braking. Most contemporary light rail vehicles have service braking rates about the same as negative acceleration. This is why emergency braking rates of about 0.3g can be difficult for standing passengers. Indeed, even seated passengers can be catapulted from their seats, unless holding on, or in a face to back seating arrangement constrained by the seat in front.

Another aspect of vehicle design is the diameter of wheels, and how the height of low floors if specified can be accommodated. Typically, tramcar wheels have a diameter of about 685 mm (27 in) which, if used in a rotating bogie, means the car floor has to be higher – for example, >750 mm (>30 in) above the rail head. For a 300 mm (12 in) floor level, the wheels would need to be less than 300 mm (12 in) diameter. This is not a practical option. So most low floor trams have fixed trucks that do not rotate with the track, and seats are arranged with spaces to accommodate wheels protruding through the floor.

Unfortunately, fixed trucks on low floor trams (Figure 2.25) create the same problems as on the first generation four-wheel trams (Lesley, 2019) (Figure 2.26). These include poor riding, excessive wheel and rail wear and the creation of rail-head corrugations. Corrugations, if not treated promptly, lead to the whole rail being ruined by this fatigue problem (Lesley, 2023). It also creates a noise nuisance of 'roaring rails' and an uncomfortable ride, although this can be partly alleviated using resilient wheels that include a spring/damping layer, which was pioneered for the PCC design developed in the 1930s.

Over 100 years ago, it was realised that rotating bogies give smoother riding and less wheel and rail wear, especially with the small radius curves, a common feature of urban systems even now. Today, light rail vehicle designers are incorporating bogies into many new low floor trams (Figure 2.27).

Figure 2.25 Some contemporary fixed truck light rail vehicles

Bombardier (Toronto)

CAF (Edinburgh)

Alstom (Paris)

Skoda (Prague)

Source: The manufacturers

Figure 2.26 Fixed truck 1930s Edinburgh tram on curve R=9 m (30 ft)

Source: Edinburgh City Council

Figure 2.27 Examples of low floor trams with bogies

Skoda (on trial)

Osaka (in service)

Siemen/ELIN (Vienna)

City Class (Birkenhead)

Rotating bogies require the inside of low floor vehicles to be very different from those with high floors, since the wheels will protrude through the floor. This can be accommodated by arranging seating units with space beneath to accommodate the wheels. This, however, limits the degree of rotation possible, so bogies need to be spaced to allow vehicles to transit the smallest radius curves. These constraints provide new challenges to the vehicle structure to maintain strength and integrity. Maximising the use of space by passengers to be comfortable and safe usually means back to back seats (Figure 2.28).

Seat boxes to accommodate wheels through the vehicle floor also means a narrow aisle, too narrow for a baby buggy or wheelchair. Special arrangements are needed in a part of the (articulated) vehicle which is easy to access, and provides adequate space for the number of wheelchairs and buggies anticipated.

Motor vehicle and automobile manufacturers spend considerable sums on market research to ensure that new vehicles are attractive and comfortable to enable maximum sales. Rail vehicle designers and system operators also need to undertake surveys of passenger and non-passenger preferences relating to the fixtures and fittings – for example, seat, hand holds, décor and lighting – to

Figure 2.28 Seat boxes to accommodate bogie wheels

ensure the best combination for the service and to optimise the attraction of passengers. In a competitive urban travel market, public transport operators and equipment suppliers need to do the same as automobile makers to maintain or increase modal share of traffic and, hence, revenue. Such attention to detail will ensure that the special needs of passengers can be met, improving the ease and convenience of light rail travel.

One way to test the potential impact of a new light rail vehicle is to build a mock up (Figure 2.29) and invite local people to provide feedback. Not only can potential passengers get a visual impression of a new vehicle, they can also test the seats, walk through, try the hand holds and provide impressions for the car building team. Mock ups have both engineering and design benefits.

Figure 2.29 Examples of new car mock ups (Trampower & Alstom)

A final and important part of the design is the power consumption of the system. There are three important variables determining power consumption

(a) vehicle weight
(b) maximum speed
(c) the number of stops.

This is because most of the energy used in running an urban tramway or light rail system goes into accelerating (kinetic energy) the mass of vehicle and passengers. On hilly systems more energy is used in climbing but can be partly recouped in descending.

The controlling kinetic energy (KE) equation is

$$KE = 0.5 \text{ x } m \text{ x } v^2 \quad (2.1)$$

where m is vehicle mass and v is the maximum speed.

A 200-passenger vehicle weighing 40 t (88 000 lb) will use twice as much power as one weighing only 20 t (44 000 lb) empty. Adding a full passenger load gives an energy saving of 40% for the lighter vehicle. Sadly, recent vehicle designs have been getting heavier. This is often due to using smaller versions of main line heavy rail rolling stock.

Reducing the maximum speed from 50 km/h (30 mph) to 30 km/h (20 mph) reduces the power consumption by over a third. This however need not result in a large reduction of service speed if higher rates of acceleration are employed.

The two graphs (Figure 2.30) show, on the left, a reduction in maximum speed leads to a service speed of 26 km/h, even with a higher rate of acceleration, compared to 35 km/h at a maximum of 50 km/h; while on the right, faster acceleration increases service speed from 34 to 45 km/h.

Figure 2.30 Comparison of different speed curves

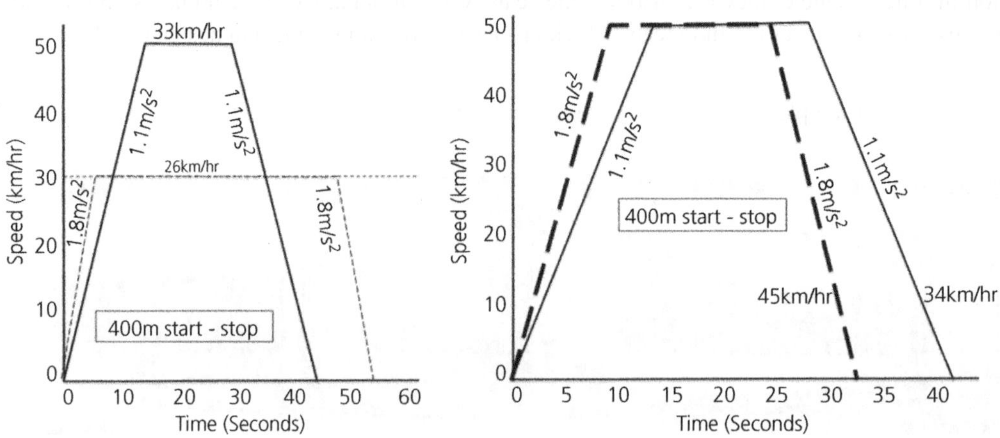

This demonstrates some of the options available to service planners to provide an acceptable journey speed for passengers. This can be tested by market research to determine the attraction to automobile users.

2.7. Conclusions

The designer of a new tramway or light rail system needs to understand the interactions between infrastructure, vehicles and operations. They cannot be taken in isolation, since one impacts on the others. It is therefore important to optimise the design of the whole system, taking into account both the first (capital) cost and the operating costs.

In most systems, the lifetime energy, maintenance and repair costs will be much greater than the first cost. So minimising power consumption, for example, can be more important than 'cutting corners' on one aspect of the construction and fitting out costs. Similarly, easy to maintain infrastructure or vehicles can significantly reduce the operating costs, with less down time needed for a smaller fleet of vehicles, or parts to repair track and so on.

A new system will be designed by a team, but the most important person is the team leader who will have a wide knowledge and experience of all the requirements for construction, equipping and maintenance. Using this, the team leader will be able to ensure that problems are minimised, particularly through co-ordination of the team to prevent clashes in designs or conflicts over construction. The main aim must be to get it right first time, since retro-repairs or alterations later are very expensive.

LJS Lesley
ISBN 978-1-83708-399-2
https://doi.org/10.1108/978-1-83708-398-520261003

Chapter 3
Designing

This chapter discusses approaches to designing and planning new systems.

3.1. Approach and demands

In designing a new tramway or light rail system, there are requirements that must be satisfied. The first is establishing the need for a new rail line. This can be in terms of the existing provision of public transport, the modal share enjoyed by public transport and, hence, the volume of private car traffic. Public sentiment can be expressed by local politicians. Next is determining the likely volume of patronage a new line will carry. From that, whether it is financially or even economically viable. The feasibility of possible rights of way will be an important consideration as well as the lines that can be constructed. The siting of tram stops must provide the maximum convenience and, so, the largest potential numbers of passengers. A thorough knowledge of the urban area and its social-economic characteristics is important. As important is an understanding of the way people behave, when there are choices of travel mode.

3.2. Setting goals and objectives

The starting point for the exercise set out in Section 3.1 is having a clear goal from which a set of objectives can be established. Since planning, including traffic and town planning, became academic disciplines, a considerable body of literature has been written on the importance of clear goal and objective setting in plan making (Figure 3.1). Confusing and sometimes mutually contradictory objectives were heavily critised in a UK light rail report (NAO, 2004). This was wide ranging with strong criticisms of tramway systems opened in England since 1990; in particular, claims by the promoters of the schemes to justify public funding. These claimed objectives were rarely met in part, let alone in full.

The more objectives a project is trying to achieve, the less likely any of them will be fully satisfied, since some may be mutually exclusive. Often in a transport or light rail project there are implicit or hidden objectives like civic 'keeping up with the Joneses' or an incumbent political party rewarding the parts of the city where its electoral strength lies, sometimes known as 'pork-barrel politics'.

What should be the objectives for light rail development? There is a body of planning literature on this topic, which is summarized in the following text.

Objectives can be ideological or determined from surveys. In the first case, the objective achieves a desired norm, like reducing air pollution to a specified healthy level or enabling people to reach their place of work within a defined time. Ideological objectives can sometimes be measured and therefore help confirm the effectiveness of the project. For a light rail system these can include carrying a specified number of passengers annually, or just in the peak period when the greatest reduction of traffic congestion might be achieved. A more ambitious objective could be to achieve a significant change of modal split, with trips attracted from cars to trams.

Figure 3.1 Plans and objectives

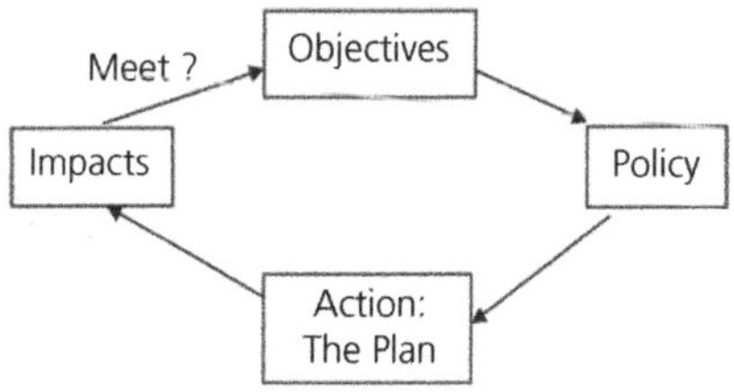

Derived objectives begin with an analysis of the existing situation to identify particular problems, but, often, subjectivity confuses such analyses. This can certainly be the case between the severity of different (subjective) attributes, even if not for the exact measure of the problem. Derived objectives can emerge from desk studies, market research, public consultation exercises or consultants' studies. Objectives like reducing traffic congestion or social exclusion fall into this category, since even defining what these 'problems' are can be difficult. Determining 'significant' changes can also be a challenge.

Measuring them may also be difficult. Identifying any statistically significant change, against what is often a noisy data background, may be almost impossible. This can degenerate into the class of the medieval debate about the number of angels that can fit on a pinhead. Nevertheless, much political debate is undertaken using this kind of language or rhetoric. The system designer will have to identify and analyse objectives that can clearly be measured and are likely to demonstrate a significant change.

Another approach is the identification of widely accepted problems and then seeking solutions. This approach can be more difficult as there is rarely a consensus as to what constitutes a problem. On the other hand, if problems can be agreed, then alternative solutions can be evaluated. This is easier than planning to achieve normative objectives. Identifying problems might be easy to achieve but not so easy to solve, since there may be alternative possible solutions. The role of the designer is then to find, examine, evaluate and select a solution most likely to resolve the accepted problems, and without creating further problems – the so-called law of unintended consequences.

It ought to be obvious that transporting people is the most important objective for a light rail scheme. In most American and European cities, the car is the main means of personal transportation, as shown in the previous chapters. A light rail scheme should, therefore, have as a main objective of changing the modal split by attracting car trips. Reducing private car use, and encouraging the use of public transport, is possible through the development of the new rail scheme. Measuring a change in modal split is normally practical, using before and after surveys. The dominance of motor car travel provides the light rail system designer with a large pool of trips from which to attract ridership, by providing links between major origins and destinations.

3.3. Transport heirarchies

Whether urban public transport operates in a regulated or deregulated market, effectively each mode of transport will be part of a heirarchy of modes, each performing a different but complementary role. This was first discussed by Willmore (1967), who observed that there is a relationship between the number (n) of modes in a hierarchy. Wilmore predicted that the exact ratio of modal speeds will impact on the speed with which journeys between every origin and destination can be completed. He proposed that a ratio (R) of 3.0 might provide the optimum journey speed for interurban travel.

$$R = V_n/V_{n-1} \quad (3.1)$$

where V_n = operating speed of mode n

and journey time (t_n) on mode n

$$t_n = k_1 V_n$$

distance of the journey (S_n) on mode n

$$S_n = k_2 V_n^2$$

where k_1 and k_2 are constants of proportionality.

From this it is possible to construct a hierarchical model (Figure 3.2), which shows how a journey from an origin can be made using different modes in the hierarchy.

Clearly in an urban context where the hierarchical modes are likely to be bus (16 km/h), light rail (24 km/h) and metro (36 km/h), pragmatically the value of R is likely to be 1.5. Further, the

Figure 3.2 Schematic transport hierarchical model

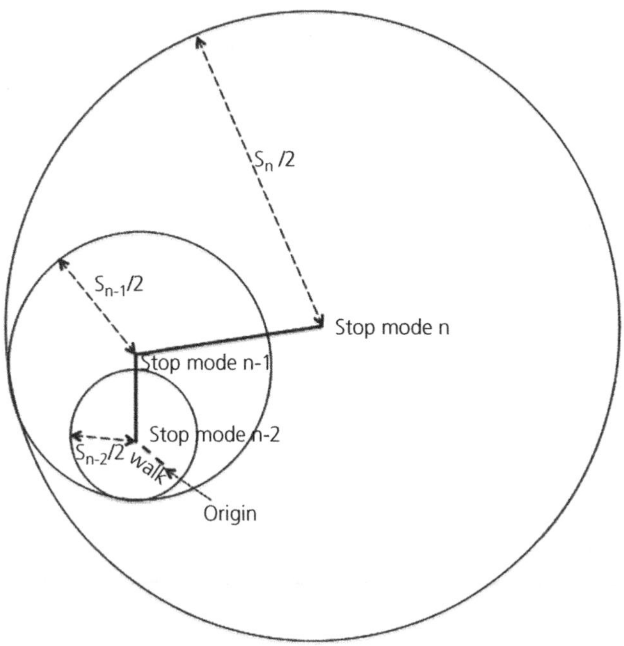

geographical size of the urban area will also impact on the number of modes that a hierarchy can support. The importance of this is shown in Table 3.1.

The number of modes operating in the hierarchy will impact on the average end to end journey times (Lesley, 1972, 1973) (Figure 3.3).

Figure 3.3 shows that for most journeys, a hierarchy with 2 modes – for example, bus and light rail – gives the fastest travel time in an urban area of 150 km^2. Compare this with Figure 1.7 for a survey of multi-modal journeys in a European city. Figure 3.3 shows a strong correlation with the actual operation of a hierarchical network. Understanding the role of different modes of (public transport) will help to increase the effectiveness of a new light rail line.

Table 3.1 Importance of mode speed ratio (R) and size of urban area

Urban area (km^2)	No. modes (R = 1.5)	No. modes (R = 3.0)
50	2	1
100	2	1
150	2	2
200	2	2
300	3	2
500	4	2
1000	4	2

Figure 3.3 Journey speed and number of hierarchy modes (area = 150 km^2)

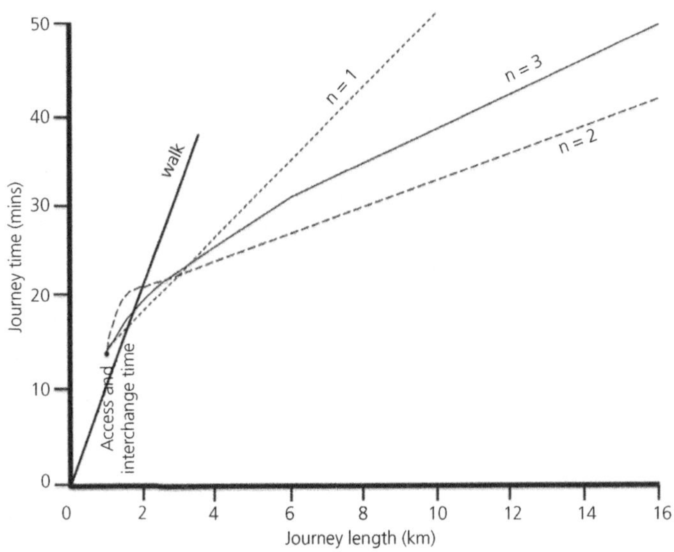

3.4. International comparisons

Light rail systems that do not carry enough passengers must surely be ridiculous? This has happened in the UK. Light rail lines have been built, which were claimed would break even on operating costs. Less than forecast patronage meant the need for a subsidy. One example would be building a line to regenerate a declining industrial area, where the population is also declining. A real-world example is the Jaén Tramway in Spain which opened in 2011 but then closed after a few weeks, when the city council audited the project and found the operating subsidy could not be afforded. The line has been in limbo since, with various plans to dismantle and sell the equipment. As of 2024 another attempt to reopen this system with a private operator is being explored, with a possible start date in 2026.

A similar dilemma occurred in Sheffield with a new three-line tramway opened in 1995. It quickly became clear that the operating subsidy was more than the local authority could afford. The network was then put out to auction and a large bus company won a 25-year concession to operate and maintain. It made significant changes to the operating practice, abolished tram stop ticket machines that were continuously vandalised and employed conductors on the vehicles. These changes raised public confidence, and increased the patronage and revenue collected by more than the cost of employing the conductors. Since then the company has produced an annual surplus of about the same amount as the cost of buying the concession. The concession was not renewed and Sheffield tramway, since 2024, is back in public authority control. The latest accounts show a net loss of £42 000 on a turnover of £25 million. By May 2025 patronage had grown to 10 million per year.

Just measuring the number of riders using a new light rail line may be inadequate or misleading. Riders may have diverted from parallel bus or other rail services and so there is no increase in total transit ridership. Establishing that riders have genuinely been attracted from private car use is also not without problems. Before and after studies of household travel behaviour can identify statistically significant modal split changes. These can be confirmed from transit rider surveys. Similarly, cordon surveys of commuters entering the central business district (CBD) can identify modal shifts, especially when linked to long term passenger and vehicle counts on the corridor benefitting from the new light rail line.

Another way to attract trips from private cars is by the use of park and ride facilities (Figure 3.4) near the edge of the urban area. This will attract external trips entering the urban area and also some

Figure 3.4 Park and ride and bus interchange at Dresden LRT

Photo Dresden Transport

trips originating in the suburbs, when a car trip to a park and ride stop may be much quicker than driving the whole way. This has a further health benefit, in that daily short walks at a park and ride stop improves passenger fitness and health.

3.5. Land use and transport integration

Transport is a derived demand from land use activity. Land use integration is a big issue, whether the local government has tight control of land use, or operates a free market philosophy. There is a strong argument that co-ordinating decisions on transport and land use development makes good economic sense. If there is a completely free market, and the land use developer and light rail system promoter might be the same, the developer will try to integrate his land use development with the introduction of a new light rail scheme, for profit maximisation reasons. Indeed, in the nineteenth century, many cities – for example, Chicago – were opened up by developers getting large tracts of land, then building rail systems, hand in hand with property development. Cities today can also adopt this approach (Figure 3.7) by linking land development with light rail extensions, providing new residents with an alternative to car use from the start.

Some city councils have discretionary powers to 'tax' new developments to contribute to transit projects. In the US, there are cities and states which have 'value recapture' taxes on property to recoup part of the rise in land values derived from new (light rail) transit systems or other public investments. In the UK public authorities can 'tax' new development proposals (e.g. from S.106 agreements in England) to pay part of the cost of new transit systems. In neither case do these charges cover the whole cost of implementing a new light rail system, especially if fares are set below an economic level and subsidy is required to cover the operating cost.

Similarly, both of these taxing approaches can be resented by developers, who have little or no say in how their money is spent on the light rail project. Developers may see a light rail scheme as a public sector toy paid for by the private sector. Developers can always look at projects in other cities with more favourable tax regimes.

Local authorities often see transport as a 'quick fix' for urban problems. This is because a new transport system can be up and running in under ten years, whereas comprehensive land redevelopment may take 50 years, the sort of time frame to which governments rarely work. Without care, a new transport system can, in fact, make an existing land use problem worse by intensifying development at critical and accessible places, and leave other areas starved of investment, suffering from 'planning blight' and becoming marginal in the urban economy. This is one of the issues discussed in 'Strong Towns' (Marohn, 2020), in trying to maintain high levels of urban 'added value' involving the resident population initiatives.

Arguing the counter factual case is just as contentious, since without any public intervention (in a new light rail system) parts of the urban economy and fabric may decline and become marginal. In urban development economic theory, this leads to a relative decline in land values, until a developer finds the redevelopment of such areas profitable. Unless of course the decline is so big that land becomes negative in value, when there is a financial incentive to disinvest.

Where new light rail schemes are promoted and part funded by property developers, the rise in land values or property prices will be of considerable interest to the developer, who will wish to see an investment in light rail rewarded. Indeed, a pioneering study of the Yonge Street Subway Line which opened in 1954 in Toronto showed that within five years, much of this transit investment had been recouped from the rise in land values and sales around station locations (Dewees, 1976) and more recently in Seoul (Bae *et al.,* 2003).

The first stage of the London Docklands light railway, which opened in 1988, was paid for within three months of construction beginning, through the rise in the value of local brownfield sites and land sales. In many places such overheated land value conditions do not apply. Declining industrial cities may effectively have land with a negative value, due to the need to remediate polluted industrial land.

The 'Strong Towns' argument is that public funds should only be employed when either the project itself generates a financial surplus to service the investment cost, or that more generally an increase in local tax revenues rewards the public investment.

Unfortunately, light rail may make little difference to this, as witnessed by the Midland Metro light rail line between Birmingham and Wolverhampton UK. This crosses large areas of industrial dereliction around Wednesbury ('the Black Country'). There has been a lack of redevelopment since opening in 1999. This means it does not carry enough passengers to cover costs. Initially these losses were cross subsidised by the local bus and tram operator. Nor, sadly, has this light rail line attracted the anticipated brownfield redevelopment or economic regeneration. This is an example of a political light rail project, in that it used an abandoned rail right of way, ignoring the fact that it hardly served any residential or work areas, but does serve a football stadium. The traffic from this, one day every two weeks is not enough to compensate for the lack of traffic on the rest of the line for the rest of the time.

Subsequently the line has been extended through the city centre to 'Five Ways' on the south side. No recent patronage figures have been published. The latest published financial statement showed that between 2021 and 2022, income rose from £7.6 million to £10.3 million, whereas operating costs rose faster from £12.5 million to £15.9 million and the annual deficit increased by 13%, funded by local taxpayers.

Objectives like urban regeneration, improving the environment, or reducing social exclusion are unlikely to result in extra revenue for the light rail operator. These and other worthy social and similar objectives, while conferring community benefits, assuming they can be detected, are unlikely to assist directly in the finances of the light rail project.

Social objectives might be considered as fortunate side effects but really should not be counted as key to justify the project. If they are, this may be the sign of a weak project. Light rail, like transport in general, is a service industry and not necessarily a driver of the economy. Cities with weak economies need to address the economy and not imagine that light rail will be a miracle regenerator. This is the case even if a property developer is promoting it as part of a package to open up access to an under used site? In this case the property developer will pay for all or part of the capital cost and offset that against the rise in the value of his land, which might otherwise not be developable.

Economic regeneration comes from the redevelopment of the land, not usually the operation of a light rail system. Transit designers must not confuse cause and effect.

3.6. Demand, economics and patronage

In order to get a new light rail line approved, there are a number of steps to undertake, which must show that there is a demand for an improvement in public transport, that sufficient patronage will be generated and that the finances stack up, making the project sensible. One of these steps includes public consultation by one means or another.

Many countries have a legal requirement for public consultation for publicly funded projects, including light rail. The reason for this is partly the democratic process to ensure that public money is wisely spent. It is also partly a 'good neighbour' exercise, to determine the effects on the residents and businesses of the area. There are formal processes in a number of countries. In the US these were set in 1976 (US DOT, 1976) (Figure 3.5).

The key to effective public consultation is stimulating feedback, both positive and negative, then addressing the issues raised. There are two main ways to achieve that. Firstly, the light rail plans will be modified to remove or at least substantially resolve any issues. If this is not practical, then explore whether or not the objectors can be compensated. While there are many examples where public consultation was successful, the cases where this failed can also be salutary.

One such example was the West London Tramway promoted by Transport for London (TfL), along the Uxbridge Road. Ironically, this was tram line 7 which was abandoned in 1938. The new tramway was to be built through the boroughs of Hammersmith, Ealing and Hillingdon. Very quickly, residents and business organised an anti-tram campaign, based on a number of objections. The objections were mostly very reasonable and revolved around the residents' access to facilities and the feared impacts on local businesses.

The campaign was led by a local resident and national journalist, Virginia Ironside. Most of the objections could have been resolved if measures and traffic management techniques in widespread

Figure 3.5 US DOT Guidance 1976

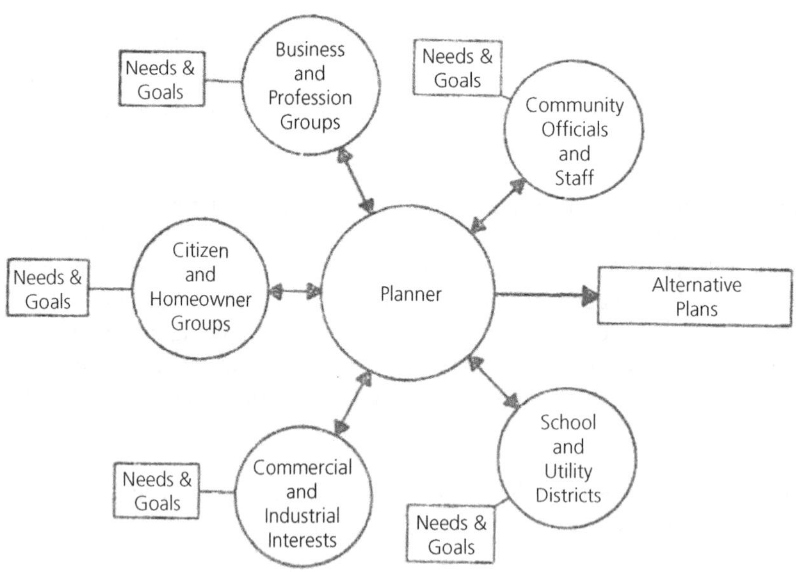

use in the rest of Europe had been proposed. Instead, TfL took a 'take it or leave it' approach. The result was that in the 2004 municipal elections, the ruling labour party lost control in Ealing and Hillingdon to the Conservatives, who had campaigned on an 'anti-tram ticket'. The Conservatives immediately withdrew support. The Labour controlled TfL had no choice but to abandon the scheme, having spent £50 million on preparations and consultation.

For new (privately funded) schemes, as well as the associated legal consultation process, the promoter will begin with local market research, to find out residents' views of their district and city. A private light rail promoter will also seek to influence public opinion. Champions will meet civic leaders to put the case for how the light rail will benefit the city.

3.7. Performance

The performance of a new light rail system will be governed by a number of factors, including

(*a*) quality of the design of the route and population density
(*b*) location and design of tram stops
(*c*) type of equipment and vehicles used
(*d*) calibre and training of staff
(*e*) operational management and timetable robustness.

There are examples of apparently similar light rail systems having very different performances in terms of passengers carried per vehicle or per member of staff, number of vehicle km run per year, or vehicle km run per member of staff. These differences have a major impact on costs and can also affect patronage and revenue, and hence viability.

An example of this was the Supertram system operating in Sheffield, England (see also 3.5). When set up, it was manned on the basis of six drivers per vehicle with off board revenue collection from ticket machines at stops. These proved unreliable and expensive to maintain when frequently vandalised. Patronage built up more slowly than forecast, making an operating loss, which the public authority was unwilling to fund. The operation was then privatised to the multinational company Stagecoach (Figure 3.6). It initially reallocated half the drivers to be conductors and fare collectors on the vehicles, abandoning off vehicle tram-stop ticket machines, which at the time cost about £20 000 each.

This immediately increased revenue by 7%, which covered the wages of the fare collectors/conductors. Having only three drivers per vehicle was very close to the Stagecoach overall average of 2.9 for its bus fleet. Increasing revenue and reducing costs meant that within two years a loss-making system was generating an operating profit of about £1 million per year, as noted by the UK National Audit Office; nearly the same sum that Stagecoach paid for a 25-year operating concession. This has now terminated and the operation has reverted in 2024 to local authority control, under the South Yorkshire Metropolitan Combine Authority (SYMCA), with a freeze in fare levels requiring an ongoing subsidy.

In most developed countries local authorities are required to produce development or zoning plans, to guide property developers as to what developments will be acceptable and where they may be undertaken. Such plans may also include proposals for transport system improvements. This can have the effect of guiding developers to particular locations where transport is likely to be improved most, or in some cases to see the transport improvement as part of the property development (Figure 3.7). The impact of these public guidance plans can be critical.

Figure 3.6 Sheffield Supertram

Source: Stagecoach Group plc.

Figure 3.7 Light rail extension and urban development, Amsterdam 1993

Photo A.P.H.Velthoen

On the other hand, a new transport line may have the opposite effect by creating ‘planning blight’, especially when the required land needs property demolition. Tenants move away. Property owners stop maintenance, pending acquisition and demolition for the new line. When this is a public sector scheme, implementation can take decades. With the relative decline in land values, land owners receive little or no compensation, when land is required for the transit scheme. This can be a source of discontentment, unless compensation is based on land values before the scheme was announced, as is the case in France.

3.8. Demand

From whichever angle a new light rail system promoter starts, good public consultation may be more than a legal requirement. It makes sense to ensure the system is acceptable to the citizens through whose ‘front room’ it will pass. As importantly, public acceptance is needed to gain the maximum ridership. For the public sector, ridership will be an important objective. For a private promoter, maximising the revenue generated will be critical to achieving the business plan projections and satisfying investors.

How many passengers will a new light rail system carry and how much revenue can be generated? The answers to these two questions are not the same, because the fare elasticity for urban (public) transport travel is rarely –1.0 and often about –0.3. In one project for a 19km light rail line, the maximum patronage was found to be at zero fares but zero revenue. Increasing fares reduced patronage (Table 3.2) but increased revenue. The effective fare elasticity is –0.17.

In urban areas where people already make trips, the purpose of those trips is known; going to work or for education, the two most important economic activities, to go shopping, visiting friends, seeking medical treatment and so on.

Similarly, the method of transport used to make these trips can also be known or easily determined. As about 80% of trips start or finish at home, the origin and destination of trips, together with the start and finish times, can be determined. This data is vital in planning a new light rail line, as it allows modal shift calculations to be made. Therefore, the likely ridership can be predicted from trips attracted from other modes and under what conditions. These calculations can be further confirmed by sensitivity analysis (Tables 3.2 and 3.4).

Looking at data (e.g. 50 years of the UK National Travel Survey), which records the travel pattern of people making trips for all the above reasons, one fact stands out. The number of trips made per capita has remained remarkably stable at nearly 900 annually. What has changed is average trip length, reflecting the real decline in the cost of travel during this period and a modal switch to private cars. Why has trip making remained stable? The principal reason is that work (or education) trips dominate most peoples' lives and normally this accounts for about two trips on most days.

$$\text{Here } n = k/c \tag{3.2}$$

where k = a constant, n = number of trips and c = cost of trip

Personal travel undertaken remains broadly static, since there are only so many activities that people can undertake in a week. What will change are origins and destinations, the trip length and the mode of transport. Since the COVID pandemic of 2020, morning and evening commuting trips have reduced in some places, as home working as been adopted by those that can. Ironically, while this has reduced the number of work trips, it increased those made for leisure. The impact on transport modes is also striking. The number of car trips in 2025 was at about 95% of the 2019 level, while bus use is only about 80%.

Table 3.2 Fare levels, patronage and revenue

Fare level (£ = GBP)	Patronage (m pa)	Revenue (£m pa)
0	6.0	0
1	5.0	5
2	4.5	9
3	4.0	12

Figure 3.8 The economic cost/demand curve

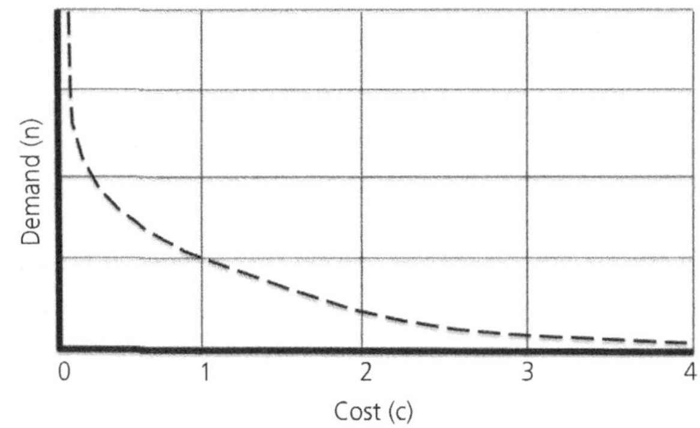

Source: The author

A light rail promoter relies on the fact that a person's mode of transport can be changed, by attracting trips from existing modes to light rail. In a light rail corridor this can initially be achieved by people in cars switching, given the right combination of service quality and fare levels. In the medium term, as people change jobs and dwellings, more may choose to locate conveniently for the light rail line, based on previous experience. In the medium term the light rail system, by enabling or perhaps providing more intensive economic developments, can also change origins and destinations, by allowing more people to live or work near or by a tram line.

As people move, some will choose to live closer to the light rail line and this will enhance adjacent house prices. In Croydon (UK) a 30% rise was reported from the opening of the Tramlink system. The enhancement of property values near light rail lines, compared to those further away, has been noticed in many cities. This property value increase provides an incentive for developers to invest in higher densities near light rail lines. This urban restructuring reinforces the role of the light rail system, which captures a larger share of urban trips, from typically 2% to over 5% of all trips per line opened, or up to 40% of the movements in that corridor.

3.8.1 Origin and destination traffic

Ridership on a new light rail line will be attracted or diverted from an existing travel pattern. Conventionally an urban area can be divided into as far as possible homogeneous traffic zones (Figure 3.9). Travel between zones will be determined either from household or similar surveys (Bonsall and O'Flaherty, 1997), or else calculated from the resident population for origin traffic and number of jobs, retail or leisure space for the destination.

Plotting the desire lines between all combinations of origin and destination zones produces a bewildering pattern for a metropolitan area of about one million people. For this reason, the data produced is recorded in an origin-destination (O-D) matrix (Bonsall 1997). While data of trips by all modes and for all purposes can be recorded in a single matrix, the light rail planner will be particularly interested in car trips for all purposes, between origins and destinations on or near the new light rail line. These will represent the largest and most attractable traffic potential.

The light rail promoter will use O-D data for trips between the zones that lie along the proposed light rail line. The other way round, the promoter will seek the zone links which have the heaviest

Figure 3.9 Urban traffic zones and desire lines from just one zone to all others

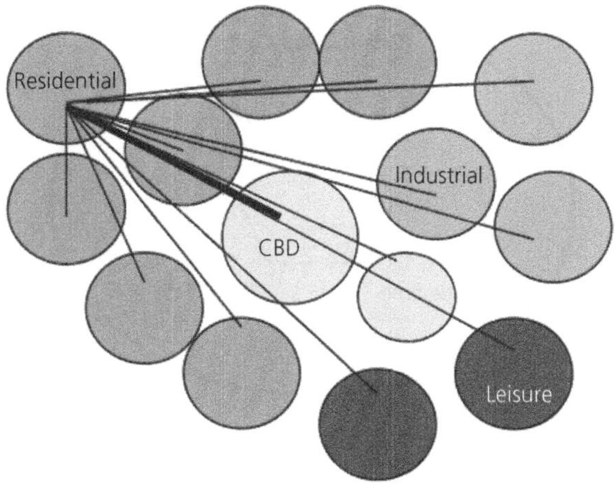

Source: The author

traffic flows. Trips from one zone to another which pass through other zones will also be of interest. In designing the light rail alignment, the promoter will seek to maximise the number of these trips that can be attracted by convenient tram stops and journey speeds.

In most urban areas there are few end-to-end trips on any particular light rail line. Most trips are between intermediate points. Therefore, knowing the origin-destination matrix data is an important foundation, both in designing the lines for a new light rail system and determining the ridership that can be attracted.

3.8.2 Generalised cost

In economic theory, the demand for products (or services) increases as the cost declines and the opposite (Figure 3.8). Analysis rests on the cost being the actual cash transaction. For more than 40 years, transport economists have known that the cash cost of travel is weakly correlated with the volume. Considerable market research demonstrates that the cash cost of transport is not a good measure of people's travel behaviour, even excluding the general elasticity for aggregate demand (Table 3.3).

Table 3.3 Typical factors taken into account on mode choice

Importance	Factor
1st	Convenience
2nd	Reliability
3rd	Speed
4th	Comfort and safety
5th	Fare

Figure 3.10 Generalised cost relationship

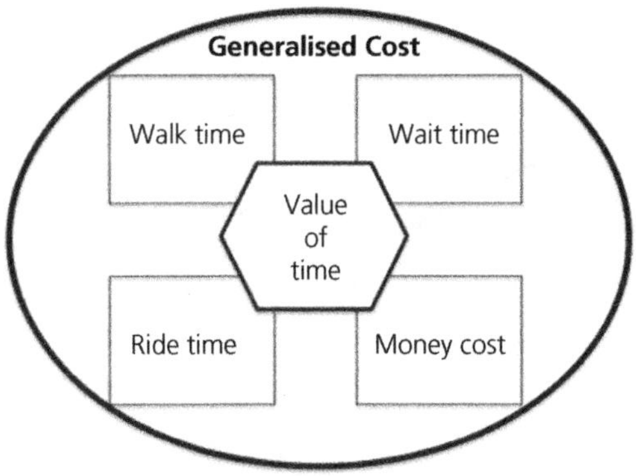

A result of this market research is that other factors were found to influence travel mode choice. These factors take into account the comparison between alternative modes of transport available for a journey. Most importantly, the quality of the journey in its entirety is considered more important than the money cost of the travel (Figure 3.10). All of these are moderated by a traveller's value of time. Bringing these different factors together was made possible by the construction of a 'generalised cost'.

Generalised cost brings together financial costs and service attributes. These are weighed against people's willingness to pay. Generalised cost has a good correlation with passenger demand. Willingness to pay is usually related to personal income and the economic importance of the journey. In broad terms, people with higher incomes are willing to spend more on transport to save travel time, buying speed. Meanwhile, those on lower incomes spend more time travelling to save money. There are many volumes devoted to the consideration of generalised cost. Here we will only consider its basic form.

The starting point is the fundamental relationship in the demand/cost curve

$$n = k/GC \tag{3.3}$$

where

n = the demand (measured in passenger km)
GC = generalised cost and
k = a constant, which is usually city specific and can be a calibration constant.

This relationship is good for specific journey purposes, so a disaggregated analysis of each journey purpose should be undertaken. This is needed to get exact relationships for the prediction of light rail patronage. Fortunately, most of the journeys made in the peak period are work related. A very acceptable approximation can be got by analysing two journey categories: work journey travel and all the rest as non-work.

The advantage of generalised cost is that it allows comparisons to be made between different modes of transport and the impact of changes of one or more aspect of a particular mode.

The generalised cost function can be written as

$$GC = c_1.T_1 + c_2.T_2 + c_3.T_3 + c_4.F + c_5.D \tag{3.4}$$

Where c_1, c_2, c_3, c_4 and c_5 are constants, and T_1 = walking time, T_2 = waiting time, T_3 = riding time, F = out of pocket costs and D = a measure of perceived comfort and safety.

From a large number of studies, c_1 and c_2 are nearly = 2. Often c_1 is about 1.8 and c_2 about 2.2, indicating the relative willingness of people to walk but a lower inclination to wait. For work journeys c_3 is usually about 1.0, meaning that people perceive riding time approximately the same as clock time. c_4 is approximately the inverse of personal wage rates for work journeys, and between 0.5 and 0.1 for non-work journeys.

People's disinclination to wait is nicely illustrated by a study undertaken by London Transport at Earl's Court Underground station in London. Here there are both lifts and escalators between deep level platforms and the street. On average the lift time is significantly shorter than escalator time. Most passengers use the escalators, because normally this can be used without a wait.

Equation 3.4 produces a value of generalised cost in terms of equivalent minutes, as opposed to one based on equivalent money. Equivalent minutes have a number of important advantages as a measure of generalised cost, since time

(*a*) is a measure which can be compared across different countries
(*b*) is (money) inflation proof, for comparisons over several years
(*c*) is good as a lower GC correlate with increased travel
(*d*) allows different income groups to be considered on the same basis.

Generalised cost can be used to determine the absolute measure of travel demand or given the existing data of demand, can be used to analyse the effect of changing the transport system as follows.

$$n_1/n_2 = kGC_2/kGC_1 \tag{3.5}$$

$$\text{So, } n_2 = n_1.GC_1/GC_2 \tag{3.6}$$

where

n_1 is the existing travel demand,
GC_1 is the existing generalised cost,
GC_2 is the generalised cost of the new situation and
n_2 is the resultant travel demand.

This has the further advantage that the calibration constants cancel, assuming they remain the same or are nearly the same during the period of change. Normally, these calculations are undertaken for changes over the next year, rather than in ten years time. So only the relative change in generalised cost needs to be determined, making predictions of changes in demand much easier to calculate, including sensitivity analyses.

3.8.3 Elasticities

Another way to determine travel demand, although normally only for aggregated data, is the use of elasticities (Figure 3.11). This is the economic measure of people's willingness to pay and can be considered as the slope of the demand cost/curve at the point being considered. For urban public transport, a fare elasticity (e) of about –0.3 has been the norm for about 40 years. This means that if fares are raised by 10% demand drops by 3%, so that total revenue increases by only about 7%. This confirms the basic point of generalised cost. The cash/fare element in the decision to travel is not the only or even most important factor.

What it also means is the assumption that the elasticity of fare increases is the same for fare reductions. So that a 10% fare decrease will lead to only a 3% patronage increase, reducing revenue by about 8%. The literature on fare decreases is much smaller than that on fare rises, since there have been fewer of them, during a period when there has been money inflation.

Elasticity e = dn/dc (3.7)

This also includes the adoption of free fares in places like Tallinn (since 2013) and Luxembourg (since 2020). Earlier in South Yorkshire, UK, (1976 – 1985) (Bly, 1976; Goodwin *et al.,* 1983) when fares were frozen, inflation reduced the real cost of public transport travel. In all cases, while there have been small patronage increases from 'captive riders making more trips', modal switches from private car travel has been insignificant (Fast Company, 2023; EU Urban mobility observatory, 2021).

There are, however, some examples of fare increases leading to more passengers, due in part to a change in the fare system and better promotion. In 1979, the newly formed Rhein-Ruhr VV in West Germany changed the fare system from a route based pay by distance scale, to a simple zonal fare system. The conurbation was divided into a number of fare zones. The cost of travel within a zone was a single fixed rate. Travel between zones was in incremental steps. This change coincided with an average fare increase of 10%.

A patronage decline of 3% had been budgeted on the basis of previous elasticity of –0.3. In fact, patronage rose by 4% giving an elasticity of +0.4. This allowed the unexpected surplus to be

Figure 3.11 Fare elasticity

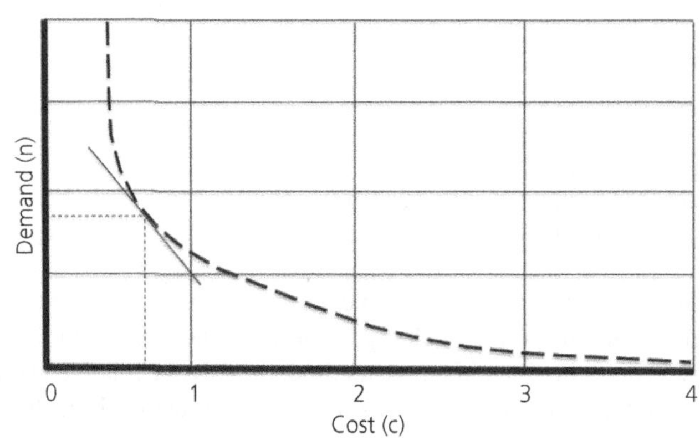

invested in service improvements, which for the next twenty years created positive feedback on ridership and revenue. One of the reasons for this counter intuitive result was that a major publicity campaign, including information at every stop and billboards, raised the profile and improved the perception of an improving system. Perhaps an object lesson in public information?

Given the level of money inflation over the last 40 years, it has not been possible to distinguish with any confidence the long run elasticity of fares. Few passengers have short-term options and cannot make immediate changes of travel behaviour. In some countries there are schemes to encourage car sharing. Where this is possible, a transit user can team up with a car driver at work, share the fuel costs and become transit independent.

The intuition is therefore that the long run elasticity is higher than the short run. Passengers without immediate choices can, over a period, make radical changes to their life style. These include buying a car, moving house to be closer to work or change jobs, as the three most obvious. These will have profound impacts on mode choice and the ridership of transit services.

Studies on the impact of new car purchases in the UK showed that the first car purchased in a household reduced public transport use by about 360 trips annually, whereas the second car purchased reduced transit use by about 280 per year (Cheek, 2008; Goodwin, 1993; Hills, 1981; Paulley *et al.*, 2004).

More recent work (Paulley *et al.*, 2004) shows that urban bus fare elasticity in the UK is now about –0.4, reflecting the fact that many bus trips are optional, as few people are dependent on buses for their journey to work. The implication of this higher elasticity is that a 10% fare rise will increase revenue by about only 5%. Bus operators cannot therefore assume that rising costs can be met by raising fares, which has been the method used by most operators in the past 50 years. This has exploited the short-term fare elasticity while ignoring the long-term elasticity. Now that the short run elasticity is higher, raising fares will merely collapse the business faster. This is the main reason that UK bus operators have concentrated upon cost reduction measures, providing useful experience for potential light rail promoters.

3.8.4 Prediction accuracy

In planning a new light rail system, the most problematic of the predictions is the patronage forecast. Indeed, the UK National Audit Office Tramway Report singled for criticism the poor accuracy of project ridership forecasts. This was in terms of not achieving the predicted patronage, resulting in lower revenue than budget. At best, forecast ridership takes significantly longer to be achieved. Both of these can lead to an unbudgeted period of losses, often being rolled up into the capital costs to be covered by the original grant or by ongoing public subsidy.

Light rail promoters normally have few options for reducing costs when opening a new system, if patronage and revenue do not reach forecasts. The infrastructure is fixed and cannot be used for anything else. Some of the rolling stock can be mothballed and staff laid off. This, however, is likely to worsen the problem. A lower frequency of service will further reduce ridership, until the ultimate situation of no service, no revenue and idle capital assets.

One way to handle the inevitable relative inaccuracy of patronage and revenue forecasts is by the use of sensitivity analysis. Here, in turn, each variable used in the prediction is altered by the same amount, usually 10% and the impact on the whole forecast determined. From this, the variable(s) most critical to the forecast can be identified.

This allows further (survey) work to refine the data and therefore improve the accuracy of the whole patronage forecast. Ranges of patronage forecasts can also be used to see how sensitive the economic appraisal model is to inaccuracies in the patronage forecasts. Table 3.4 shows the impact of changes to the operating speed, from a base of 20 km/h to 25 km/h. Raising the operating speed by 25% increases the patronage by nearly 1% and the revenue by 2%. While increasing speed does not itself significantly improve patronage or revenue, it will reduce operating costs by about 25%, needing fewer vehicles and staff. A 25% reduction in operating costs can make the difference between viability and losses.

A major variable in forecasting patronage is the degree of competition that a new rail line will face from car and taxi use and buses in places not regulated. Unregulated buses have been the case in much of the UK since 1986 and many developing countries. If the urban transport system is integrated then normally bus services will be recast to support and feed a new rail line. In a deregulated environment, light railway transport (LRT) must offer a competitive service to attract a viable market share. Where new LRT systems have been opened in a competitive urban travel market, achieving patronage forecasts can take longer.

Two UK examples show that such competition may not be fatal. In Manchester when the Metrolink opened the first line between Bury and the city centre, a bus company began an express service at less than the tram fare. This lasted less than a year, as not enough bus passengers were attracted from the trams. The second was in Sheffield when the line between Meadowhall and the city centre opened, a local bus company began an express service using articulated vehicles and at half the tram fare. This only lasted six months.

A new LRT system can exploit car travel by means of strategic park and ride stops. These can increase the capital cost of a new system but, by attracting trips that might otherwise not use the system, revenue increases. The local community benefits from a reduction of car trips and congestion on city roads. This benefit can be enhanced by the use of a central area parking policy.

Table 3.4 Sensitivity analysis to different operating speeds

Speed	20 km/hr	Base	22 km/hr	Faster	25 km/hr	Fastest
Line	Pax (m pa.)	Fares $m pa	Pax (m pa.)	Fares $m per year	Pax (m pa.)	Fares $m per year
1	0.73	2.9	0.74	2.9	0.74	3.0
2	1.29	5.1	1.29	5.2	1.30	5.2
3	0.69	2.7	0.69	2.8	0.70	2.8
4	0.20	0.8	0.21	0.8	0.21	0.8
5	0.65	2.6	0.65	2.6	0.65	2.6
6	0.88	3.5	0.88	3.5	0.88	3.5
7	0.50	2.0	0.50	2.0	0.50	2.0
Total	**4.94**	**19.6**	**4.96**	**19.8**	**4.98**	**19.9**

None of these techniques by themselves will make patronage and, therefore, revenue forecasting 100% reliable but at least the likely range of forecasts can be narrowed. A business tool like minimax analysis can be used to minimise the worse-case scenario. In a commercial project, the investors are likely to consider only the low estimate of ridership if this is adequate to finance the capital, then, should demand be higher, investors will enjoy comfortable profits. Publicly funded projects appear to have used only upper forecasts, if the UK National Audit Office conclusions are valid.

3.9. Stops

Passengers access light rail systems at tram stops. These are the first physical point of contact with the service. They are the 'shop window' of the system and bring the system into perspective. Whatever marketing image has been created, will be reinforced or damaged by stops. Functional and aesthetically pleasing stops need not be expensive and, because they have low running and maintenance costs, can be a better investment than, say, further lines or more vehicles.

Tram stops need to be located giving convenient access to where people live, work, shop, go to school and so on. Based on behavioural studies, people will walk up to 400 m to a transit stop, unless urban development is not even. This figure is based on a number of studies. In Melbourne, Australia, the distance was found to be 500 m (Taylor, 2024).These figures imply a stop spacing of between 400 and 1000 m, with a continuous corridor at least 800 m wide served for walk and ride passengers. With park and ride, or bike and ride in some countries, stations can be further apart without reducing the effective population catchment. Longer stop distances mean a faster service speed and shorter travel times. This, however, can be moderated using a mixture of compulsory and request stops. The latter being those with an expected lower level of usage. The timetable can then be constructed using the probability of halting at request stops. Ideally, stops should be located near road junctions to maximise the catchment area on the basis of central place theory (Christaller, 1933) Integrating stops with pre-emptive traffic management measures can improve vehicle flows and reduce congestion.

There is an interaction between stop spacing and operating speed. Fewer stops mean a higher operating speed but smaller passenger catchment areas. More tram stops means a slower speed but shorter times spent at each stop to set down or pick up passengers. It also means that passengers have a shorter walk to reach the nearest stop. As walk time has a worse generalised cost than riding in a tram, there is a fine balance over the distance between tram stops.

Like the supermarket queue, stop dwell time is a critical factor in maintaining service regularity. One of the main advantages of light rail is that stops have a relatively low capital cost and can be quickly built, even in streets. Ideally, stop platforms should be on a straight section of track making boarding and alighting fast and safe. Typically, stop times will be between 10 and 15 s, which is an important part of route planning. Where there is no demand, but future development will create traffic flows, a space can be reserved for a future stop.

Having assessed the likely potential traffic generation of each stop, planning a new light rail system should concentrate on the stops with the greatest potential traffic and be prepared to add new stops later, as revenue is generated, to increase patronage further. This is a decision that can only be taken based on excellent local market intelligence and co-operation with the local planning authority.

Adding new stops later has advantages. Firstly, once a system is open, operating performance should improve and, therefore, higher operating speeds can be achieved. New stops can often be

Figure 3.12 Tram stop in Kaufhof Department Store, Ludwigshafen

Source: SWR Aktuell

added without needing new rolling stock, by using the extra efficiency gained, while maintaining or increasing the operating speed. Secondly, in terms of the local travel market, it allows the operator to be seen to be developing the business, by opening another 'outlet' for more customers. Thirdly, it allows the operator to undertake joint ventures with property developers, on land adjacent to the line, opening new stops integrated with the development. Indeed, electric light rail can pass through buildings without any special environmental protection, as occurred in Ludwigshafen, Germany, where a department store was developed with a tram stop inside (Figure 3.12).

Light rail stops are the shop window of the system. So it makes good sense to have high quality designs and finishes. It is a strategic decision whether to have a standardised design for all stops, or local design for each stop reflecting the character of the neighbourhood. Whichever approach is adopted, a corporate identity and clear signing helps passengers find and be guided around the light rail system. Good design will reduce any resistance from local residents to having a light rail stop as a neighbour, unless it takes away peoples' front gardens. This was a factor which led to the demise of plans in the early 1990s for a light rail system in Birmingham, UK.

A new light rail system should operate at a high frequency, to minimise waiting times, by offering a 'turn up and go' service. Therefore, the need for stop facilities is minimised. Short and reliable waiting times are crucial to maximising the attraction of trips from cars. Personal security is also important. Good CCTV coverage is essential, but the evidence shows that, more effective than CCTV, is a busy station on a well-used system, where waiting passengers provide social pressure and cohesion, make travel both pleasant and give riders a feeling of safety.

3.10. Co-ordination with property development

The best way to integrate land use and transport is to have property developers as investors in the new light rail system. This gives the developers a financial incentive both for the investment in transit and to undertake complementary real estate developments. These will mutually enhance the light system patronage as well as making the new property more attractive, needing less car parking and, therefore, the potential for more or larger properties. This increases the total returns to the investors. Some cities are specifying the maximum number of car parking spaces per new dwelling. In London the maximum figure is one space for two dwellings, and at some locations zero spaces.

It also means that with better access, the land value will rise. This avoids what is seen as dead money, through the taxing approach discussed in 3.5. Mobilising one or more developers works better with a genuine partnership between the public and private sectors. Although there are many examples in the US and in continental Europe, sadly, to date, there are few such partnerships in Britain.

Even if there is no direct relationship with property developers, a new light rail line will have an impact on the property market. Research in Germany (Hass-Klau, 2004) showed that property prices in the catchment area of a new light rail rose between 9 and 19%, compared to similar properties further away. In Britain, the picture was more mixed, but light rail can attract premium prices. In Croydon this was found to be 10%, in Manchester 12% and Nottingham 15%.

3.11. Meeting civic and environmental objectives

There is no doubt that a new light rail system can significantly raise the public image of a city and its perception by outsiders. One measure of this is an analysis of tourist postcards. In cities like Amsterdam, Toronto and Zurich, the tramway is a major feature of these cities' postcards. London's double deck red buses are similarly iconic worldwide. Where, increasingly, retail shops and other organisations are branches of international companies, the unique personality that light rail confers on cities can be an important civic icon. It is also under the control of local people. This is important for reinforcing city-hood and can be measured by the everyday use of the system. Even in cities with high car ownership, light rail systems will be used by virtually everyone, with the usual distribution between frequency of use and percentage of population.

Most developed countries are now committed to reducing greenhouse gas emissions under the Kyoto Treaty and the subsequent Paris and later United Nations COP agreements. A 'zero carbon' future will be helped by electrically powered light rail, using renewable generation. As well as reducing carbon dioxide and other gas emissions, there will be an important contribution to city environments, including reducing the levels of vehicle traffic, congestion and noise. This is especially so if car trips are attracted to tram. With electricity renewably generated, light rail systems will directly reduce fossil fuel consumption and greenhouse gas emissions.

Light rail emits no air pollution at the point of operation in city streets. It can therefore improve the air quality of the districts through which it passes. This is particularly true for the health threatening emissions PM10 and NOx. These emissions are particularly connected with respiratory and cardio-vascular diseases. In the UK traffic pollution is estimated to kill about 45 000 people annually, compared to under 2000 in road traffic crashes. Zero emission light rail can operate through sensitive areas (Figure 3.13), like pedestrian zones.

Additionally, light rail is significantly quieter than internal combustion engine transport. Even electrically powered road vehicles still generate noise from rubber tyres on the road, typically about 70 dB(A). Urban noise pollution has been cited as a problem by World Health Organisation (WHO). This is both a health problem and one that degrades the urban (social) environment. The WHO recommends a maximum level of 55 dBA external day time noise. This compares to many urban streets with noise levels over 70 dBA, which makes conversation difficult or impossible.

There are many examples of light rail vehicles operating through pedestrianised precincts (e.g. Figure 3.13), where other modes of transport would be unacceptable or anti-social. It is difficult to value the ability of light rail to improve both civic image and prestige, and at the same time offer real improvements in the quality of life for citizens. Light rail into urban centres provides fast travel and a high level of accessibility. This helps to reinforce the role of central areas. Trams are also good neighbours.

Unfortunately, none of these aspects will be reflected directly in the finances of the system, but they can be important influences in obtaining the required municipal support and legal approvals. It can also be important in the vital business of attracting inward investment for industry, commerce or housing, where a city's image may be the major factor in attracting such investors. This has proved significant for Freiburg which nearly doubled its population and more than doubled its gross value added (GVA) over 20 years, thanks to an expanding tramway. Freiburg Figure 3.14 now claims to be Germany's 'greenest city' (Litterbins, 2020).

All the above considerations provide strong justification for the promoters of light rail to undertake full and genuine public consultation, as discussed in Section 3.3. A city inheriting a new light rail system, which has not involved the demolition of property or the disruption of communities during construction, will provide willing supporters and enthusiastic riders.

Figure 3.13 Light rail in Freiburg pedestrian streets

Source: John Krause, Clerk Street Project, Chicago

Figure 3.14 Trams in Freiburg

Sources: Freiburg Tramways and Tony Young

3.12. Freight on light rail

Historically, many light rail systems carried freight, from parcels on streetcars to dedicated freight cars. Some current light rail systems also carry freight. This is separate from railways/railroads through streets dedicated for railway traffic. At the heaviest end, light rail can share railroad tracks, as in San Diego, where regular locomotive hauled freight trains operate during the night when the light rail service is suspended. For many years, some European countries used city streetcar tracks to move one or more rail freight wagons, between mainline sidings, factories and distribution depots. Usually, small electric locomotives haul the wagons (Figure 3.15).

More recently, old tramcars have been converted (Anderson, 2009) to carry dedicated freight containers – for example, in Dresden (Streeter, 2009). Here CARGO trams are operated for Volkswagen, to link three factories in the city with a merry-go-round service using the streetcar network carrying pallets of parts. This saves an estimated 200 truck movements per day (Figure 3.15). Earlier, Glasgow's tram tracks were used by local ship yards to move materials and parts needed for building ships. Zurich has a fleet of cargo trams, at first only to move garbage, but now a wider range of goods. Finally, Amsterdam has converted old passengers trams to carry freight.

Carrying freight in secure wheeled cages is also practical with low floor light rail vehicles (LRV) and level boarding (Figure 3.16). This has been introduced by a postal company in Germany to

Figure 3.15 Examples of cargo trams

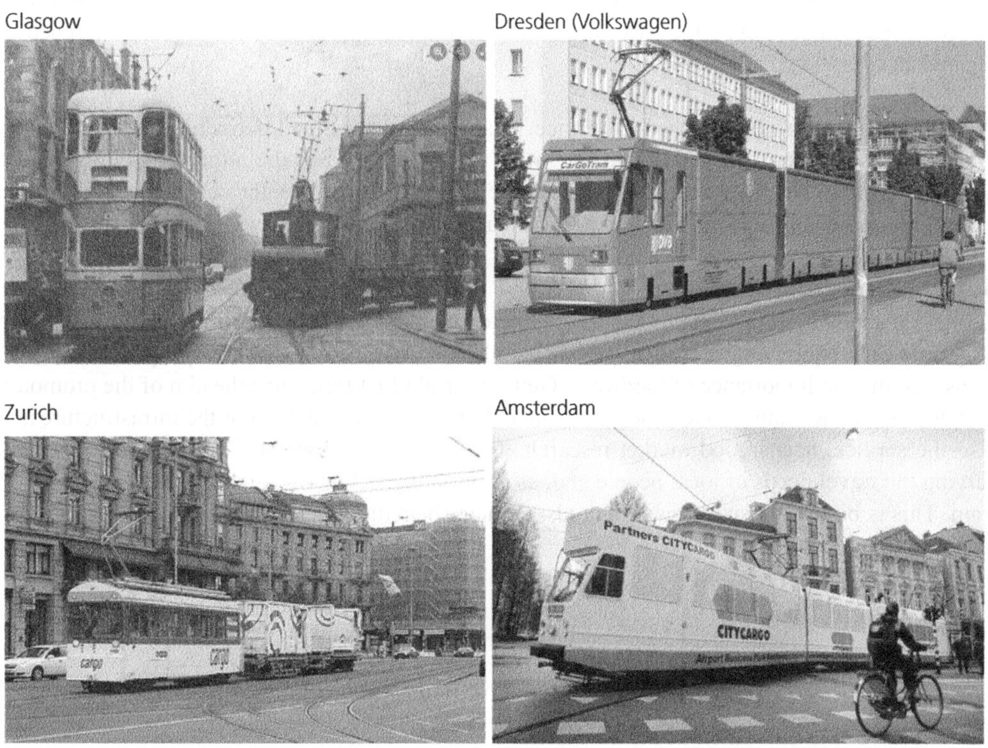

Sources: Scottish Museums, JRug Collection, www.thetransportpolitic.com and www.parool.nl,

Figure 3.16 Wheeled freight cages by tram

distribute mail and parcels within a city. Such movements would also allow local shops to be serviced without the need for trucks parking in the street, saving traffic congestion and air pollution. Such a freight service could be combined with consolidation centres. Clearly, careful planning of access to any freight depots is required, as is access where wheeled cages will be loaded and unloaded. Integrating freight movements into a high frequency passenger service can be a scheduling and an operational challenge.

Wheeled pallets can also be used to service local shops, with timed deliveries and the manual movement from stop to shop. Moving freight by tramway increases the utility of a city's system and provides another income. It also reduces the number of road freight vehicles operating and gives environmental benefits from less pollution and noise, and freer flowing traffic.

3.13. Conclusions

Designing a new light rail line, or even a completely new system, is a multi-disciplinary task. Not only does this require engineering expertise but also an understanding of how people make travel decisions and the importance of hardware. Getting it right first time must the aim of the promoter, both to save money and also to produce revenue. Public acceptance, both of the infrastructure and also the service, needs good market research. The rewards are big for a well-designed system, serving the travel needs of local people and, as importantly, achieving a modal switch from car to tram. This is both a transport reward but also a major benefit to the city from reduced congestion and pollution.

LJS Lesley
ISBN 978-1-83708-399-2
https://doi.org/10.1108/978-1-83708-398-520261004

Chapter 4
Engineering preparation

This chapter discusses the engineering problems of designing new systems and possible solutions.

4.1. Introduction

Translating the design stage discussed in Chapter 3 into a project that can be constructed, operated, maintained and developed needs considerable engineering inputs. These involve not just the conversion of the design drawings into engineering drawings but also the modification of the highway pavement so that light rail vehicles can be safely accommodated with other traffic. Engineering drawings allow the light rail system to be constructed quickly, safely and right first time.

Light rail has very different demands compared to heavy rail. Operating speeds are lower, the vehicles lighter, the curves and gradients more demanding and, when tracks are embedded in roads, there is the interaction both with the road pavement and traffic and the impact of heavy road vehicles on the tracks.

Light rail demands the interaction of three major engineering disciplines: civil to design the structures and tracks, mechanical to design the vehicles, and electrical to design the electrical system to power the vehicles and the traction and other electrical systems within the vehicles. Of these only electrical engineers have a common interface both with civil and mechanical engineers, over the design of the infrastructure power and vehicle traction equipment.

Civil engineers must interface with highway and traffic engineers over the use of street pavements for the track, with mechanical engineers over the rail/wheel interface. This includes the resilience of the track supporting structure and the suspension of the vehicle body on the wheels. Track quality also impacts the performance of the pantograph, or other current collector, in contacting the overhead line (OHL) needed to provide the power to the vehicle for traction and auxiliary systems. Electrical and mechanical engineers must interface over the traction motor torque and transmission to the bogies and wheels.

These interfaces are not new, and there is considerable experience on ways to resolve problems for a new scheme. Incremental developments from earlier projects can obviously be applied where the new system is merely using previous technical solutions. Engineering advances, with new ideas and approaches to both materials used and designs adopted, will require testing and evaluation. Where these advances reduce costs or improve service, the promoter will have an incentive to assist in such technical developments. Innovations can open new ways of doing things but also have a risk. Someone needs to understand the innovation and have the courage to try new ideas.

Compared to 40 years ago, the production of engineering drawings and testing has been made much easier by the use of computer-based analysis tools. These can be applied to all aspects of the engineering design. Computer-aided design (CAD) is now a mature area and packages like AutoCad, Inventor and SOLIDWORKS allow designers to undertake detailed three-dimensional designs for their aspects of the project, but also to file share with other engineers. This allows a proper consideration of integrating structural and dynamic interaction between the infrastructure, vehicles and equipment and to avoid at least physical clashes for space.

Modelling from satellite measurements and ground penetrating radar (GPR) allows the civil engineer to examine many options for the alignment and installation of the light rail tracks. At the same time, the civil engineering team can consider the cost, ease of installation and the minimisation of traffic and urban disruption during construction.

Mechanical engineers can use finite element analysis to ensure the strength and robustness of body designs and dynamic simulation packages to model the behaviour and performance of running gear and the vehicles. The electrical engineers can model power consumption and current flows, dependent upon the vehicle design and operating proposals. At station locations, acceleration power needs are dependent on likely passenger loadings and the topography of the line. This allows the design, capacity and location of substations to be optimised, as well as the feeders for sections of lines.

These however are only applicable where the new designs are an interpolation from earlier experience. Where new designs require an extrapolation of earlier work, then invariably physical testing may be needed to confirm and calibrate computer models. Even then, the experience of light rail vehicle body fatigue problems (Hondius, 2007) shows that only physical testing (Figure 4.1) can finally confirm the durability of designs, even when there is an assumed measure of over design.

Figure 4.1 Trampower wheel testing 1994

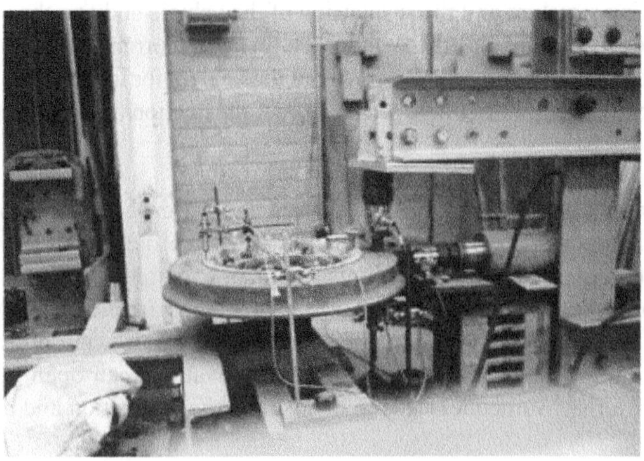

Source: Ralph Boese

4.2. Specifications and standards

A new light rail system design should ensure that all interfaces are compatible. This is to avoid creating problems. Solving these could be worse than the original problem. Resolving a problem at the design stage should be the aim of the design team, since this is the best time – the least disruptive and costly opportunity.

A well-versed project engineer will lead the team of civil, mechanical and electrical engineers. The project engineer will identify problems early and then use resources to avoid or mitigate them. The easiest option for new light rail systems would be to clone a previous system, which is clearly working satisfactory. This would be fine if every city was identical, with the same set of geographical problems to be considered. Unfortunately, this is never the case, as no two cities are the same. Similarly, travel patterns and origins and destination distributions will differ.

The 'cloning' approach was the one adopted by the new generation of light rail systems in North America opened from 1980, where equipment was obtained nearly off the shelf from Germany. This had the advantage of both previous extensive system testing and experience, and a range of manufacturers to offer proven equipment. Especially important was the Frankfurt U2 car developed in a partnership between the Frankfurt Tramway Operator (FVV) and the Düwag car company of Dusseldorf. Düwag had a proven mono-motor traction bogie (Lesley 2025) on which light rail vehicles are mounted. Nearly all the new systems, including Calgary, Edmonton, San Diego and Sacramento, adopted this design. Indeed, Siemens, which acquired Düwag, opened an assembly plant in California to supply the North American market. Later, the new Supertrams for Sheffield were mounted on Düwag mono-motor bogies.

These first new systems also adopted German designs standards (BöStrab) for tracks, power systems, and signalling and control. Modifying USA railroad standards, proven Germany technology could be used with little or no modification. While the technology may be easy to clone for a new system, more inventive solutions will be needed for other problems. These are usually network or geographical challenges that are unique to a particular city. Without solutions to those problems, a new light rail system cannot economically be introduced. Sometimes as in the USA, changing the regulations may be the key to introducing innovation.

The transit industry must move on, just as the automotive industry is doing in changing from internal combustion engines to battery electric. Innovation is also needed in response to the requirement to give equal access to all passengers, including those with disabilities. The biggest challenge is to accommodate passengers in wheelchairs. The first-generation new American light rail systems adopted high floor vehicles. Retrofitting these for wheelchair access under ADA was expensive and led to the consideration and introduction of low-floor vehicles *c.*300 mm (12 in) to top of rail (TOR).

The main reason to use low-floor LRVs is that it is difficult and expensive to retrofit accessible high platform stations in city streets, to provide level boarding. This issue was also faced by the Manchester Metrolink which opted for high-floor vehicles in 1991 and profiled high platforms in city streets to give level boarding with ramped access from the street level, as is also the case in the Hannover example (Figure 4.2).

The International Union of Public Transport (UITP) has over many decades produced standards and guidance over the design and construction of light rail systems. It is also working with the

Figure 4.2 High platform in street of Hannover, Germany

Source: Hannover Tramway

European Union (EU) towards the creation of standards for all areas of transit industry and activity, including light rail. Until that is achieved, there are differing national standards and approaches to the approval of light rail systems. As the USA has shown, this does not stop the use of other standards. In the EU, approval to one country's standard should be reciprocally recognised in all other EU countries. In practice, such inter recognition is far from complete.

The UK used to have a set of tramway standards but abandoned them when tramways were closed throughout the country, except for Blackpool.

An EU wide set of light rail standards will inevitably be heavily influenced by German practice (BöStrab), since Germany has the largest number of operating tramway and light rail systems, and nearly the longest continuous experience. EU standards will also need to address the difficult task of enabling innovation, without the need to subsequently rewrite the standards. This is one of the strengths of the German (BöStrab) approach, which is performance based. This should mean EU standards are based on performance, delivery and impact, rather than the more usual prescriptive technical standards used in many countries.

4.3. Constraints

For implementation engineers, the main constraint is to take a design scheme and make it work and be fit for purpose. Inevitably, compromises are required. In the introduction to this chapter, there was a discussion on the interfaces between engineering disciplines in the design of light rail systems. There are more and probably more difficult interfaces between the light rail system and the rest of the city, in terms of civic expectations, architectural appreciation, traffic and highway management, as well as being an acceptable neighbour to residents.

The issue of under street utility plant has already been discussed (Section 2.4.1). There are also interfaces with highway and traffic engineers over the structure of the roads being used for light rail lines. This also applies to the way that traffic is managed during system construction, which may be particularly important in minimising congestion and traffic delays. Traffic management after the light rail system has opened is also important. This is especially in relation to any reallocation of road space. There will be interfaces between the system electrical engineers and electricity supply companies over the required power and location of substations to feed the light rail network.

Aesthetic and environmental considerations may place a large constraint on the installation and implementation of a light rail system in historic city centres. The intrusion of OHLs to supply the power to LRVs in some areas may be contentious for new systems. There are, however, many systems with OHL which are long established, even through sensitive historic areas, with World Heritage status – Figure 4.3 shows OHL fixed to and supported by the Vienna Opera House.

The visual sensitivity of OHL was one of the reasons for the development in Bordeaux of yet another surface contact system to supply power to the vehicles (Figure 4.4). As in previous attempts, over a period of more than 100 years, there have been considerable difficulties. The principle of the Bordeaux 'third rail' system is that a plate in the road is energised when the LRV passes over. This is achieved by a radio signal and a current collecting shoe under the tram runs over and contacts the plate.

This surface contact system costs about 400% more than OHL. The Bordeaux system was developed by Alstom. Bombardier has also developed a power supply system that does not need OHL. Alternating currents are induced in electrical coils set in the roadway between the running rails. Under the LRV is an induction coil, the second part of a transformer using an air gap for the magnetic induction. At the time of writing no commercial service is operating with this system. All such systems are confined to central areas, with changeover points to the outer/suburban sections using OHL. However, the advance in both the capacity and cost of batteries is making this a possible alternative.

Figure 4.3 Vienna OHL and Opera House, World Heritage Site

Source: Eurocheapo

Figure 4.4 Bordeaux tramway, without OHL using third-rail surface contact

Source: Frauscher Technology

Previously a third-rail 'conduit' system used slots in the road and underground conductor rails. This was used in several capital cities including Washington DC (Figure 4.9) but was expensive to install and maintain. More recently, OHL less systems have been developed using on-board battery storage. These, however, have limited hill climbing and accelerating power, even when combined with regeneration on braking. Batteries increase the mass of vehicles, requiring more energy for acceleration and increase the cost, with little saving compared to OHL. Heavier vehicles wear out the track faster.

The architectural dictum of form following function is particularly true for new light rail systems in urban areas. One obvious example is how to support OHL, if used, by traction poles or building suspension? Establishing the engineering parameters for the OHL to have the right strength and durability is not difficult. Typically, the tension in span wires is about 1200 N, which most masonry walls can easily accommodate (Figure 4.5). If a more costly system, like that adopted in Bordeaux, is used, this will also complicate the design of the tracks.

The OHL designer should attempt to ensure that the equipment fits well into the urban environment. Two schools of thought prevail. One is a pastiche of earlier architectural styles, to mimic historic city centres. The other is contemporary designs, so that the system has a clear character and cannot be confused with other street furniture.

Unfortunately, there are examples where neither of the above have been adopted and brutally functional installations used, like the 'I' beam traction poles in the streets of Croydon (Figure 4.6) and Manchester. These would be more fitting on a private right of way or a mainline railway electrification. OHL should be integrated into the other street furniture, including sharing lighting standards, where possible.

What applies to OHL, also applies to track installation, station design, street equipment (e.g. electrical junction boxes) and, of course, the light rail vehicles themselves. Often vehicle manufacturers

Figure 4.5 Examples of OHL building mountings: 1900 and 2020

Figure 4.6 Croydon heavy-weight OHL West Croydon Station

Source: Google Streetview

will build a mock-up of the proposed LRV (Figure 4.7) and show it off, inviting feedback both from the public but especially from potential riders and system promoters. It is also important to get ergonomic feedback from the people who will operate the system, since they will be working for many hours inside the vehicles.

Ergonomics relates primarily to comfort and safety. It is not difficult to design LRVs that are comfortable and safe for passengers and crew, since there are plenty of examples of successful and

Figure 4.7 Examples of vehicle mockup

Sources: Trampower Ltd and Transport for London

some less successful designs. Ergonomic standards, however, need to be interpreted by industrial designers to achieve an aesthetically pleasing finish and one that is easy to construct and maintain.

Earlier light rail engineers were able to achieve harmonious designs, which were efficient, functional and complemented the urban scene. New light rail design engineers have a heavy responsibility not to distract from cherished city streets and thereby increase public resistance to new systems being opened. Pleasing form that satisfies the operating function will win over citizens as passengers.

4.4. Tracks

While there are systems that claim to be 'trackless' (Section 1.3), all vehicles need a support structure to operate. Without a running surface, light rail systems will not work. Tracks confer three important advantages

(a) automatic steering
(b) low rolling resistance, an order of magnitude less than rubber tyres on roads
(c) return path for the electric current taken from OHL.

In addition to these, tracks must safely support light rail vehicles, transmitting static and dynamic forces into the ground. In streets, tracks also absorb the forces imposed by rubber wheeled (heavy road) vehicles. Tracks also define the path that LRVs will follow. In pedestrian areas this is a big safety advantage. The same can said across complex highway junctions, with suitable lane markings. Indeed, evidence shows that tramways will discipline traffic, reducing conflicts and improving safety.

4.4.1 Automatic steering

Unlike mainline railways, light rail systems have tighter radius curves often as low as 10 m (30 ft). On railways, wheels steer through the forces created by the conicity of tyres and inward inclination of rails. This method of steering however only applies to curves with radii exceeding 200 m (600 ft). For many light rail curves, most points and turnouts have radii about 25 m (80 ft) or less. Here vehicles are steered by wheel flange contact against the rail side and flange check rails, one of the causes of wheel squeal and rail wear (Figure 4.8). This has important impacts on track design and maintenance.

Figure 4.8 Worn side rails on Blackpool track curve 2006

Source: The author

4.4.2 Rolling resistance

The coefficient of friction (and adhesion) of steel wheels on steel rails lies between 0.1 and 0.3 depending on track, weather and wheel conditions. Rubber tyres on roads have a higher coefficient of friction of about 1.5, added to which pneumatic tyres flex as they roll, absorbing further energy and heating the wheel. The low rolling resistance of light rail vehicles can be calculated from the Davis (1926) formula (Equation 4.1)

$$R = 6.5 + 320/wn + 0.046v + 0.096av^2/wn \tag{4.1}$$

where
w = max axle load (t)
n = number of axles
a = frontal area (m^2)
v = maximum speed (m/s)

For a six-axle tram with a maximum axle load of 10 t and maximum speed of 50 km/h (14 m/s), the power needed to overcome rolling resistance is given by:

$$R = 1200\,N \text{ at } 14\,m/s = 16.8\,kW$$

Low rolling resistance means that most of the power consumed is for accelerating the vehicle mass and this can be determined from Equation 4.2.

$$KE = 0.5\,mv^2 \tag{4.2}$$

where
KE = kinetic energy (Joules)
m = total mass (kg)
v = maximum speed (m/s)

Thus, the tram shown above for the Davis formula (Equations 4.1 and 4.2), in accelerating, will consume 3 kWh at a rate of 108 kW, which is about seven times more than the rolling resistance, giving a total power needed for acceleration of 125 kW.

This is an important design parameter for a new light rail system to help optimise power consumption, since frequent starting and stopping are central to urban passenger operation. One way to achieve this is by lighter vehicles. This will reduce acceleration (kinetic) energy needed and also overcoming rolling resistance. As a comparison, a double deck bus with a passenger capacity of 80, weighs about 12 t empty (about 150kg/passenger space). Typical light rail vehicles for 200 weighs about 40 t empty (about 200kg per passenger space).

If they had the same weight per passenger, the LRV weight would be 24 t. Ironically buses use about the same energy per passenger as light rail, because the lower rolling resistance of LRV is lost by the greater kinetic energy needed to accelerate a heavier LRV. Diesel buses of course need oil, while LRVs use electricity, which can be generated from a variety of fuels, including renewable sources.

4.4.3 Return current

In light rail systems the tracks are part of the electrical circuit between the vehicle and substation. This was not always the case. The Washington DC system (Figure 4.9) until 1962 used an underground current supply system known as the 'conduit'. This had a slot between the running rails, through which a pickup passed. The rails had no function in the traction current circuit.

In comparison to the total electrical resistance of the traction circuit between substation and LRV, the rails are at least an order of magnitude lower than the OHL. This means that provided the rails are fully bonded electrically, there will be no stray currents. This is discussed more fully in Section 4.6.

Figure 4.9 Washington DC, underground conduit current supply

Photo Joe Testagrose

There are two classes of light rail tracks

(a) embedded in highway pavements and shared with traffic
(b) segregated for exclusive use of rail vehicles.

4.4.4 Embedded tacks

Light rail tracks in streets are normally laid with the rail head horizontal and level with the road surface, with the rail web vertical. This minimises the impact of rubber wheeled road vehicles on the rails. Heavy road vehicles are normally more damaging than light rail vehicles. Tracks embedded in highways need to be able to transmit the static and dynamic forces from the rail head into the subbase. This is more fully discussed in 'Embedded rail tracks' (Lesley, 2025).

Grooved girder rails are about 180 mm (7 in) high. This usually means a foundation of about 300–500 mm (12 in–20 in) deep below the rails as support. The rails also need to be kept to gauge, which for most of the history of tramways has been achieved by bars (tie rods) bolted through the web of the rails, allowing for lateral adjustment. Sometimes sleepers or concrete blocks can be used instead to which the rails are directly fixed.

Switches and crossings or turnouts will normally be a special design for the particular street location and junctions, to fit in with the street topography and to protect the footway at the side of the carriageway. These are normally prefabricated, with three main units (Figure 4.10), two the same (switch unit). Plain rails will join the three units to make a complete turnout for a single track. For a double track turnout, ten units (four switches and six crossing) are needed. Prior to shipping to the installation site, the complete fabrication will be assembled in the factory to ensure that it matches the detailed track drawings and so can be reassembled at the (road) junction correctly.

The highway provides the foundation for streetcar tracks. For this to be successful, the pavement and its subbase must be capable of carrying the static and dynamic loadings of light rail vehicles (Figure 4.11). If the pavement is already carrying heavy road vehicles with similar axle loadings,

Figure 4.10 Turnout units

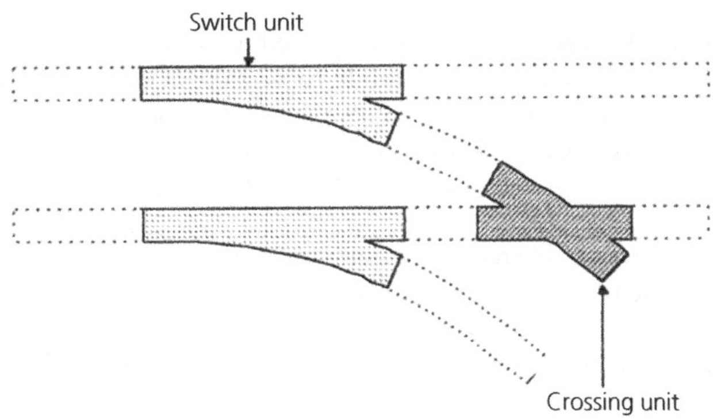

Source: The author

Figure 4.11 Typical section of a flexible highway pavement

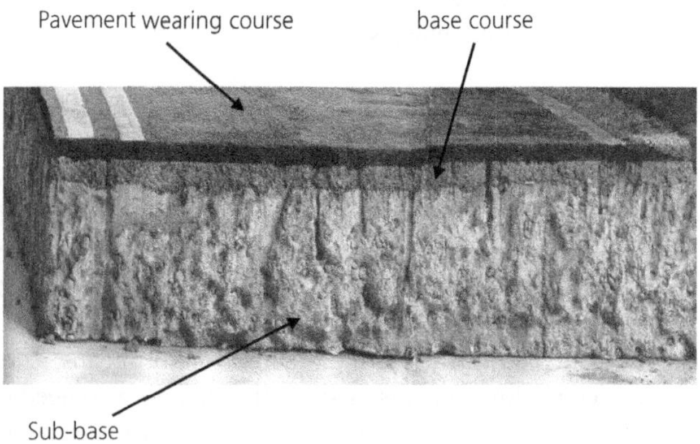

then there should be few problems. Pavement design is also well understood and rests upon the suitability of the ground under the road. The universally accepted measure of ground strength is the Californian Bearing Ratio (CBR). The lower the CBR, the thicker the pavement has to be for a given weight and volume of traffic, with maximum axle loads and expected total life (Atkins, 1980) (Figure 4.12).

The pavement thickness of a downtown or arterial road carrying 30 000 or more vehicles per day will be between 300 and 380mm (12–15 in), depending on the CBR. Provided there are no surface signs of pavement failure – for example, rutting (fatigue), potholes (subsidence) or wearing course needing replacing – then the subbase will have a CBR more than adequate to support light rail or tramway tracks.

The track robustness is needed to accommodate both LRVs and HGVs, whether using orthodox methods, with a monolithic concrete slab (Figure 4.13), or newer less intrusive designs (Figure 4.14), where the strength of the road base is used as the foundation for the track. This consists of a separate pre-fabricated concrete longitudinal beam for each rail, only 180 mm (7 in) deep and 360 mm (14 in) wide (Lesley, 2002).

Both methods require the road pavement to be excavated to allow rails to be laid. The slab track method is more intrusive, takes longer and usually requires under-street utilities to be relocated away from the footprint of the new track slab, since they will be inaccessible. The design and coordination of utility diversions is worthy of a text in its own right. Usually, the utility companies can undertake relocation but need to be paid and scheduled into the track installation programme.

Closing streets to make this possible requires careful traffic management, as well as steps to maintain access required for premises along the road. The construction programme approach adopted – for example, a block-by-block installation or the whole downtown area in one installation – will determine how comprehensive and long lasting the disruption will be.

Ideally, traffic rerouted for track installation should be the same as the ultimate traffic management regime, when the light rail system is operating. This will get drivers used to alternative routes. It will then make the light rail system simpler to adopt.

Figure 4.12 CBR, pavement thickness and traffic

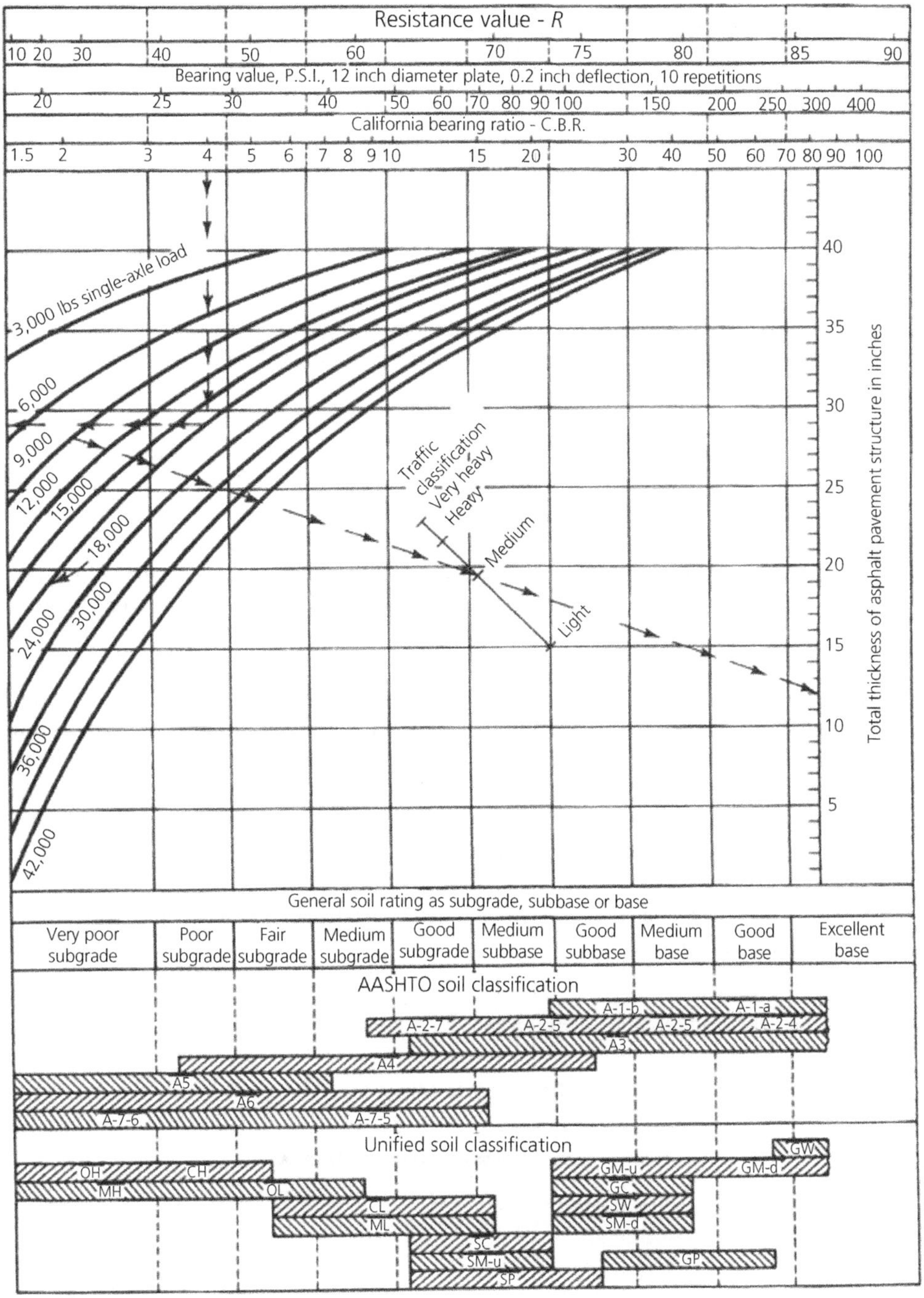

Source: Atkins 1980

Once cleared of traffic, excavating the road is a major undertaking, even if some of the spoil, which has a hazardous nature but can be reused on site, has to be stored elsewhere to create a working space. Clearing away the road pavement can be undertaken with backhoes of similar excavators. Excavating the utilities, unless these have already been decommissioned, is nearly always by hand. If the utilities have been decommissioned, then any plant below the foundation slab level can be left in situ.

Backfilling and compacting the road base in readiness for casting the concrete slab has to be done in stages, with spoil laid in 75 mm (3 in) layers and compacted to a CBR of at least 5%. The track (support slab) is cast in situ upon this, which can be done with temporary shuttering in sections or slip-form paved with a one-pass machine. This can include, as in Sheffield, continuous slots to take the rails.

When the concrete is cured, the rails are welded into long strings, for electrical continuity and then clipped to the slab. Finally, the pavement running surface is replaced up to the level of the top of the rails (Figure 4.13).

The success of this method relies on the rails being rigidly fixed to the concrete foundation slab, usually by rag bolts or similar clips, and that the surfacing forms a watertight seal to the rails, to prevent water penetrating into the pavement. In practice, over time the fixings work loose and heavy road vehicles rock the rails and break the seal with the pavement. Once broken, surface water is then pumped into the pavement, from passing heavy vehicle wheels, further damaging the installation. This leads ultimately to the failure of the pavement or the rail fixings, or, in the worst case, both.

A failed pavement will not only give an uncomfortable ride for road vehicles but, with potholing, will be potentially dangerous. There will be special arrangements in different cities as to who is responsible for the maintenance of the highway pavement around light rail tracks. The light rail operator will, however, always be responsible for maintaining the tracks in a safe condition. From experience, across Europe's some 300 light rail systems, it is heavy road vehicles that are the principal cause of damage to embedded street tracks.

For this reason, a new embedded track system was developed to exploit the existing road pavement strength, rather than having to rebuild the highway structure, and also to avoid the need to relocate

Figure 4.13 Girder rail street embedded tracks

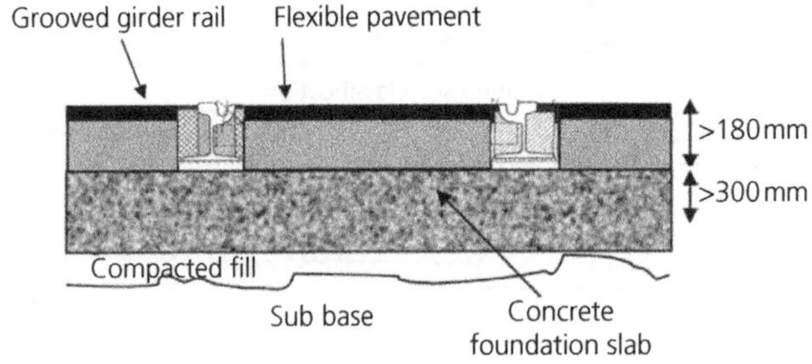

Source: The author

Figure 4.14 LR55 Strip beam foundation track

Sample LR55, and with transition rail in Sheffield

Source: The author

under street utilities. The LR55 track system (Figure 4.14) separates the structural functions of a girder rail. LR55 rails use the vertical strength to support concrete foundation beams, which are very stiff vertically and horizontally. The LR55 rail retains the strength for LRV steel wheel support and guidance functions.

This results in a shallower and lighter rail and, overall, a much less intrusive installation. The key to making this work is the use of a polymer to bond the foundation beam into the road pavement and rail into the concrete beam. The polymer will be strong enough to resist thermal forces in the rail. It also provides an elastic support structure for the same. This has the benefit of reducing rail head damage, like corrugations, to almost nil and gives a smoother ride for rail vehicles and their passengers.

The LR55 system (Figure 4.14) is also more robust for heavy road vehicles like buses, as it is more resistant to the kind of damage that girder rails with thin check rails suffer and does not jolt vehicles crossing or going along the tracks.

The LR55 track is a 'mass-spring-mass-spring' structure, that attenuates ground bourn vibrations by about 30 dB. This also reduces noise (at 100 Hz by *c.* 20 dBA and throughout the audio range by 10 dBA). The LR55 track installation is also simpler and quicker, requiring only a minimum of pavement excavation, under 10% of that needed for girder rails with a slab foundation. Using prefabricated concrete beams means that a section of road need be closed for about a week while about 100 m (110 yd) of track is laid. With practice, a track gang could lay up to 100 m per 8 h shift. Additionally, one rail at a time can be laid. This means that no road needs to be fully closed during track installation, with temporary contra-flow traffic management allowing traffic to continue to flow. An animation video has been prepared to show this process (Wondertram, n.d.).

As part of the testing programme, the LR55 track system was installed in Rotherham bus station (Figure 4.15), with 2 500 bus impacts per day. During a 2-year test period, over 2 million buses impacted the rail. This was the equivalent 30 years of the bus traffic on a busy European urban road. After this the LR55 track was used to replace a failed section of 80 lb girder rails on the Sheffield Super Tramway.

Figure 4.15 Rotherham bus station LR55 installation inspection

Source: The author

The smoother head of the LR55 rail means that heavy road vehicles crossing the tracks do not face a sharp upstand of the groove check rail. In the UK, this has led to the rails being damaged, potentially exposing light rail vehicles to the danger of derailment (Figure 4.16). The Sheffield installation of the LR55 track replaced an '80 lb' girder rail, that had failed in less than six months from heavy goods vehicle impacts.

The LR55 track was installed over a weekend in March 1996 and has remained in continuous service, needing no maintenance since. Furthermore, with the continuous elastic support, light rail

Figure 4.16 Track damage and loss of check rail in Manchester 2009

Source: The author

vehicles on the LR55 section of track are quieter and no corrugations have formed (Lesley, 2009). The LR55 rail heads show little sign of wear, and the rails should last a minimum of 40 and probably over 50 years without attention or needing replacement. Finally, the polyurethane bonding is not delaminating despite over 300 heavy road vehicles (44 t gross vehicle weight (GVW)) crossing daily.

The LR55 track system had previously been comprehensively tested (Lesley, 2002). This included cyclic testing at main line axle loadings (25 t with 200 million cycles) and up to 80 t axle loads. Failure tests over voids needed a 59 t axle load and pull out (30 t) many times more than thermal expansion forces. Testing was conducted under water and in temperatures between −10°C and +60°C. Electrical resistivity testing was undertaken in laboratory conditions and in the field. Live testing was undertaken at Rotherham bus station, the track was instrumented, data logged and analysed. On the Sheffield Super Tramway, there are over 5 million t of light rail vehicles and about 4 million t of heavy road vehicles annually.

In the absence of a different approach to embedded track design and installation, damage from heavy road vehicles is going to be a permanent, disruptive and expensive maintenance cost. This also ignores the higher first cost of installation. The example of the 3.4 km (2 miles) street tracks in Manchester, installed in 1991 over 52 weeks, cost £8 million (US $13 million) to relocate under street utilities and £6 million (US $10 million) for the track installation. As well as regular patch and repair during the summer of 2009, these rails were replaced at an undisclosed cost. The estimate for laying the Manchester city centre tracks with LR55 was £7 million and using only one track gang (not three) would have taken just 34 weeks.

4.4.5 Segregated rights of way

The most economic construction for light rail is on segregated rights of way as a classical rail track, with rails clipped to ties (sleepers) on a crushed rock ballast (Figure 4.17). Some early pioneers of segregated light rail rights of way – for example, Winnipeg and Liverpool – softened the visual impact of 'railway' tracks using grass (Figure 4.18). Unfortunately, the grass retained rainwater, which rotted wooden sleepers. More recently, German designs, using concrete units to support the rails, allowed grass to grow without shortening track life or increasing the maintenance costs (Figure 4.20).

Figure 4.17 Schematic of construction of segregated tracks (see also Figure 4.26)

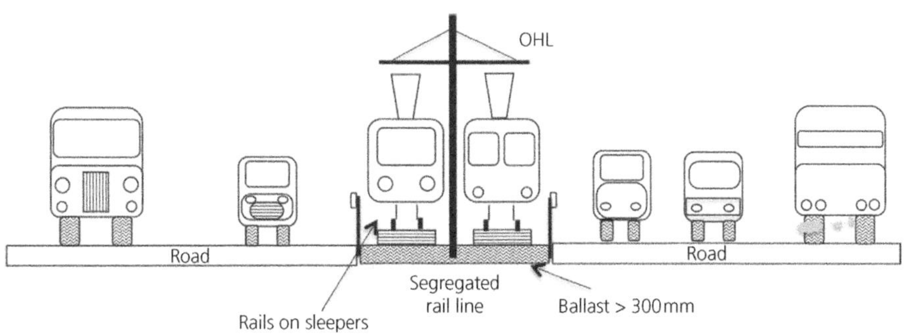

Source: The author

Segregated rights of way are often in the middle of dual carriageway roads but can also be found alongside roads and completely away from roads, if older or abandoned rights of way – for example, canals (e.g. New Jersey) or railways (e.g. Manchester and San Diego) can be used. The cost of creating new rights of way in dense urban areas, in buying property, rehousing residents and businesses, and demolition is very high, so is best avoided if possible.

While segregated light rail rights of way using the median of divided highways was pioneered in North America and England in 1914 (Horne and Maund, 1982), it was taken up on a city wide basis in Liverpool. The visionary city engineer, John Brodie, planned the expansion of the city based on 37 m (120 ft) wide radial and ring roads. These new roads were all built with a grassed median strip 10 m (33 ft) wide. This approach cost much less than half the cost of tracks in highway pavements. The first segregated light rail line was opened in 1914, using a clinker ballast base to support wooden sleepers (ties) 2.74 m × 0.25 m × 0.15 m (9 ft × 10 in × 6 in) and infilled with clinker. Soil with grass seed sown was laid on top to just under the rail head. Sadly, there was inadequate drainage, which led to the sleepers rotting prematurely.

The benefits of segregated tracks led to this idea being taken up by cities in Europe rebuilding from after Second World War damage, as a low-cost way to improve public transport. Tracks were laid separate from the road carriageways in new reservations (Figure 4.19). This improved both the economics and service of tramways. Low costs extensions were built into new suburban residential areas, leading to improvements in the engineering of segregated tracks and reduction of capital and operating costs.

The principal engineering improvements involved better prepared track bed and grading to ensure that rainwater fully drained away. Special drains were built where the ground could not absorb rainfall. Larger sized (>25 mm/1 in) aggregate in the ballast improved drainage. Concrete sleepers

Figure 4.18 Segregated grass tracks, Liverpool 1932

Source: GN Southerden

now nearly predominate, although in the spring of 2009, wooden sleepers were still being used for street track renewals in Budapest.

Finally, on segregated tracks, conventional (grooved or groove less) rails are clipped either directly or by way of a base plate to the sleeper, with the rails long welded to give a smooth ride and electrical continuity. In this case, as the rails are exposed to the weather, precautions have to be taken to lay the rails at a neutral temperature, or compensated by tensioning the rails, to prevent buckling in hot weather.

Where new roads are being constructed, a median strip wide enough for light rail can be provided at a marginal cost and certainly for less than building a new right of way. In the near future, few major urban dual carriageway roads are likely to be built, so segregation of light rail tracks will have to exploit existing roads. This can be achieved by re-allocating carriageway space on the basis of the capacity provided (Figure 4.19). A typical light rail system operating on a five-minute frequency with 200 passenger capacity vehicles will have a peak capacity of 2400 passengers per lane per hour.

The same space on a six lane freeway will have a vehicle capacity of 1500 cars per hour, at a peak period average occupancy of 1.2 passengers per car, meaning a maximum flow of 1800 people per hour. On a radial city street, a maximum flow of about 800 cars per hour can be expected, giving a passenger capacity of under 1000 per hour. A light rail gives a 33% capacity increase compared to a multi-lane freeway and over 50% better than a four-lane city street. Converting one of two road lanes to light rail improves the peak passenger capacity from 900 to 2850 per hour, a 300% improvement, without the need for property demolition and only basic traffic management.

Figure 4.19 Road lane reservation, Szczecin, Poland

Source: Neil Pulling

This approach was pioneered in Amsterdam, where other traffic was diverted from tram tracks. Such 'Trambaans' also have another positive benefit of providing congestion free routes for the emergency services to reach incidents quickly, saving time and lives. Figure 4.19 shows an example from Poland where only white lines 150 mm (6 in) wide separate light rail vehicles from general traffic on an inner city street.

A frequent light rail service using such surface reservations is normally self-enforcing, without needing legal penalties. In the peak period, light rail vehicles passing stationary queues of congested cars is a good incentive to switch modes. At least at the margins, enough car trips will use light rail, reducing congestion according to Mogridge (1990), whose thesis is that increasing the speed of public transport is the only way to reduce traffic congestion and car dependency. Having easy park and ride is another key to making this successful.

New light rail systems built out to the suburban parts of a city can use rights of way separate from highway traffic. These should be close enough to residential areas to make the service convenient for residents living nearby. Figure 4.20 shows a new residential area in Freiburg where light rail was built simultaneously with the urban build out. Figure 4.21 is an example of such a new line in Madrid, where the light rail line is alongside a highway. Grassed tracks soften the visual impact of tracks, as most of the time when there is not a light rail vehicle passing, it looks like a park. This also helps to absorb some of the carbon dioxide and other fumes from the adjacent road traffic, as found in research (Kappis *et al.*, 2016).

The light rail engineer can now demonstrate how a new urban transport facility can be a friend of the Earth and also a friend of the system funder, by economical designs. Infrastructure that has a long life and low maintenance cost will be the cornerstone of a viable light rail system. More importantly, the engineer, in building a convenient and comfortable line with a high frequency and reliable service, will maximise the ridership. Attracting trips and reducing the volume of motor car traffic benefits the whole community.

Figure 4.20 New residential area in Freiburg

Figure 4.21 Lawn light rail tracks in Madrid

4.4.6 Rail problems

Track construction includes different combinations and methods in most light rail systems, reflecting various rights of ways. The action of steel wheels running over steel rails creates wear and failure problems, of which the track engineer must be aware. With regular maintenance, rails on straight tracks should last at least 40 years. Curved tracks have shorter lives which, if not maintained, can lead to less comfortable and possibly less safe running. On curves, vehicles will grind the vertical face of the (inner) rails and create corrugations, a fatigue failure (Figure 4.22), and also put the gauge out of tolerance, allowing the wheels to climb up the rail face onto the rail top. The Nadal equation (Dukkipati, 2000) will predict this condition and allow preventative maintenance.

Figure 4.22 Corrugated damaged curved track

Source: The author

Track or wheel lubrication can help to minimise rail side wear but not eliminate it. For this, the regular reprofiling of the rail is required, and ensuring that the wheel profile is within tolerance. Reducing rail side wear should also include using the check rail of the inner radius curve, to share lateral (centrifugal) forces. Grinding the rail side can put the track out of gauge. Where plain rails are used, then swapping the inside and outside rails, to use the unworn rail sides will extend the life of the track. Ultimately rails will have to be replaced;on severe curves typically between every 8 and 10 years. Swapping over grooved rails is not possible.

Fatigue is a likely problem for track engineers. The most common of these is the creation of corrugations on the rail head (Lesley, 2009). These are caused by unstable running along the rail head, creating work hardened zones at a regular interval (Figure 4.22). While this is not usually a safety critical failure, it generates uncomfortable riding inside the vehicles that can damage the vehicle suspension. Corrugations also generate noise, which is a nuisance to those living or working alongside the tracks. This phenomenon is known as 'roaring rails'.

Eliminating the formation of corrugations requires wheels and rails to be kept exactly to profile and identical diameters to reduce unstable running, but, once formed, only rail head grinding is a possible cure. The grinding process has to remove about 2 mm (0.1 in) to prevent the roots of the corrugations reaching the bottom of the rail head. Unless this is done regularly, only rail replacement will remove corrugations. Regular grinding of the railhead also helps to maintain a profile for more stable running and, therefore, is less likely to lead to the formation of corrugations. This remedial action does however shorten rail life.

Another area where the design engineer will need to pay attention is turnouts, switches and crossings. This will be particularly so if there is uneven light rail traffic between the branches over the crossing (Figure 4.23). The nose of the crossing will be subject to differential wear and damage. The nature of the damage will be determined by whether a single casting has been used or the crossing has been fabricated (Figure 4.24). Fabricated crossings are also more vulnerable to damage from heavy road vehicles. This can be minimised in the track alignment planning by avoiding the need for heavy road vehicles to pass over the crossing.

Where road traffic is heavy, trams can switch tracks before junctions, using 'pre-sorting' turnouts and 'sleeved' tracks. This means that problems of track switches being damaged by heavy road

Figure 4.23 Mosley Street, Manchester, and a pre-sorting junction in Sheffield

vehicles can be removed from the problem area (e.g. Mosley Street in Manchester) compared to Sheffield (Figure 4.23).

Damage to street track crossings will also be affected by whether the groove is raised through the crossing and LRV wheels pass over on their flanges, not their tyres. If wheels pass over on their tyres, the nose of the crossing will be impacted and subject to damage by wheels. This is unlikely where wheels pass over the crossing on their flanges. Compare the wear of a cast crossing in Sheffield to a similar-aged fabricated crossing (Figure 4.24).

Figure 4.24 'Silent' casting and fabricated 'railway' crossing

Source: The author

4.5. Stops and stations

Stops and stations perform two important functions. Firstly, they provide the interface between passengers wanting to travel and the vehicles providing the transport. The second function is to provide a timetable observance place, so that the service can be reliable and regular. The crucial design and engineering question is how many stations and where should they be located? This also impacts on the acceptability to the local community.

The academic and engineering literature has many station optimisation papers (Lesley, 1973, 1998). Most however are based on uniform density catchment areas (e.g. Lesley, 1976), which few cities possess. These papers are only a guide to actual station spacing and location. In practice, identifying station sites require considerable local knowledge and research. As well as needing to provide an economic and safe service, stops need to maximise the potential patronage.

Assessing the local travel market is the other approach. How far are people prepared to walk to reach a light rail or tramway stop? Here the literature is clearer at about 500 m and sometimes up to 800 m. In a catchment area of 800 m radius, the average walking distance to the station will be about 530 m. When generalised cost assessments of walking are considered, the light rail designer seeks to minimise walking distance, since this is perceived to be nearly twice as irksome as in-vehicle time. More stops or closer location of stops improves passenger accessibility. Walking times to a stop should be less than 15 mins (*c.* 1 km/1094 yds). This then is a trade-off between reducing walking time, having a slower service speed and minimising the overall door to door time.

This is one of many design conflicts. To make walking distances shorter, stops need to be close together. When stops are close together, the average LRV service speed is reduced, if trams always stop at every station, lengthening the in-vehicle time might make the overall trip less attractive unless the origin – destination time is reduced. Furthermore, the slower the operating speed, the more vehicles are needed to provide a given service frequency. This increases the need for staff and so overall costs.

Generalised costs and market research exercises show that waiting at stops is even more irksome than walking to and from stops. For a reliable and regular service, the acceptable waiting time can be shown to be half the service frequency interval. So the operator has a large cost and patronage incentive to achieve a high service speed. This maximises vehicle productivity and generates or attracts the maximum patronage and hence revenue.

When putting all these factors together, the light rail system engineer must consider stop spacings of about 500 m, which gives an average walk of 340 m, about 4 mins and high operating speeds of about 25 km/h. The key, however, is a careful assessment of local conditions, since in suburban and low density residential areas, longer stop spacings may be both possible and desirable – for example, two recent extensions in Freiburg (Figure 4.25) have different stop spacings; in Vauban stops are about 400 m apart, whereas in Rieselfeld they are 600 m. If park and ride (both car and bike) is considered, this reduces access times to light rail stations, even when further apart, with a higher service speed, so shorter overall journey times.

In the more densely developed inner and central parts of a city, stops can be as close as 250 m (Figure 4.26), especially if the route alignment includes curves where vehicles must in any case slow down. Shorter stop spacing will not significantly reduce the operating speed on that part of the line. It is then practical to offer stops more convenient to important traffic attractors or generators, like shopping centres, universities, hospitals, cinemas, office complexes and so on, than would be the case for a subway or metro service. Accessibility will be much better, therefore generating more patronage, which has been proven on many light rail systems.

The final point on station spacing and location relates to the network attributes.

Busy highway intersections are good places for tram stops, maximising the catchment and alternative available routes, as well as meeting safety and other requirements. In accessibility terms,

Figure 4.25 Stops in Rieselfeld and Vauban, Freiburg

Figure 4.26 Light rail stop, Central Nottingham

Source: The author

highway intersections are the most connected parts of the network. Christaller's 'Central Place Theory' (1933) shows that high value activities locate to the most accessible points in transport networks. For the light rail system designer, therefore, network nodes are valuable places for stops, with the maximum accessibility to and from the local catchment area, and the potential for sympathetic property development.

The stop design will depend on the geographical and topological conditions, and what, if any, traffic priority the light rail system enjoys over other traffic. Ideally, an approaching LRV should get a pre-empted traffic signal 'green' and a clear run into or out of the station. Therefore, only the time to stop and set down or pick up passengers is needed. The maximum service speed can therefore be maintained.

The stops and stations are the permanent public interface of a light rail system, with the vehicles as a mobile face. Ideally, intending passengers should only wait a minimum time, at the most a couple of minutes, for the next departure. A 'turn up and go' service is the most attractive in getting people to switch from cars. This wait time should also be in comfort and safety, with shelter from the weather as appropriate. Reliable service information increases passenger confidence. Passengers need both reassurance and guidance on how best to use the service, especially if new to public transport.

New systems will need to pay particular attention to this, especially to attract car commuters. With a frequent service, waiting times will be short, so elaborate seated waiting rooms are not needed. In cities with high rainfall, shelter for a peak accumulation of passengers will be needed (Figure 4.26). In cities with hot weather and overhead day time sun, shade will also be important for the comfort of passengers, which can be provided by trees (Figure 4.27).

As the most permanent and visible part of a light rail system, stops need to be good neighbours. Local residents will not appreciate loud public address announcements late at night. The design is both part of the corporate image of the system and functionally important to attract and retain

Figure 4.27 San Jose low platform station at Virginia on Highway 87

Source: Nathan Cook

riders. Considerable design effort should therefore be expended to build stations that enhance the local neighbourhood, and so be adopted by residents as part of their civic infrastructure.

As well as attracting passengers, stops are also where passengers alight to reach their destinations. Here navigation from the stop must be clear and simple, especially if it includes interchanges with other modes of transport – for example, buses and taxis. Some stops may also have park and ride, so passengers must be able to find their car or bicycle easily. Passengers must also leave their vehicles in complete confidence that they will be there on return, undamaged. Designing the parking area needs to use the principles of defensible space and not just rely on security systems like CCTV. These have dubious records of preventing damage or theft. Constant and large volumes of people circulating is the best deterrent.

Defensible space includes real people, whether passengers passing through, operating staff on duty, or neighbours having a clear view of the area. Any person with criminal intentions must know this, since being seen or caught by a real person is the largest deterrent against crime.

All of the above used to be achieved by light rail staff on duty to provide security and passenger safety. Very few light rail systems can now afford such dedicated operational staff. That is not to say that people cannot be employed, especially where heavy passenger flows are expected. Franchises can be offered for ancillary services like newspaper/magazine sales, coffee shop, fast food and so on. These would be a way in which passengers can see a regular person. This can also help to form part of the reassuring relationship that is needed to turn car commuters into transit users and to make a success of a light rail system.

A light rail station franchisee might also perform some minor system functions, on an agreed contractual basis, at much less cost than a dedicated member of staff or a roving staff team. This can include an alert system for security, if there is a problem at the station, or to call the emergency services should there be a passenger accident. Franchisees can also sell light rail tickets, provide service information and even keep the platform area clean.

The above focused on the functional needs of stops to attract and satisfy passengers. There are more prosaic needs in designing and building stops. These revolve around the kind of vehicles that the system will use, the likely peak volume of passengers and how they will reach the stop. For new light rail systems, it is likely that vehicles will be partly or completely low floor, giving level boarding from platforms about 300 mm (12 in) high, compared to metro platforms which are 900 mm (3 ft) or more high. To make this easy for all passengers to board and alight safely, platforms need to be level with the doorway thresholds.

There has to be some tolerance with this, since vehicle wheels and rails wear and the relative position of the floor/door threshold will change. Some LRVs have adjustable suspensions that can accommodate wear. It is usual to allow some tolerance, typically as much as 50 mm (2 in) has been found from ergonomic studies as the maximum safe vertical difference between platform and car floor that elderly people and those in wheelchairs can safely manage. With a platform of 250 mm (10 in) height, cars with floors 300 mm (12 in) from the top of the rail (TOR) can operate safely. This is also the maximum tolerance under various disabled access regulations, with typically a maximum 50 mm (2 in) horizontal tolerance (Figure 4.28). As wheels and rails wear, the vertical height difference will decrease.

Platform length will depend on vehicle length and whether, at some stage, coupled LRVs may be operated. The most prudent approach would be to build to the required system specification on opening. This will accommodate the longest car to be operated, with space reserved for extensions, either required by increased passenger numbers from the initial system or the addition of new lines sharing tracks and stations. There will always be an economic calculation comparing extending platforms to running LRVs more frequently.

The minimum platform width is normally dependent on regulatory requirements. A new system should aim to provide wide platforms, to give a sense of openness and safety. Typically, a free-standing platform will be at least 2000 mm (6 ft 6 in) wide, ideally 3000 mm (10 ft). Integrating

Figure 4.28 Light rail station considerations (see also Figures 4.17, 4.24 and 4.25)

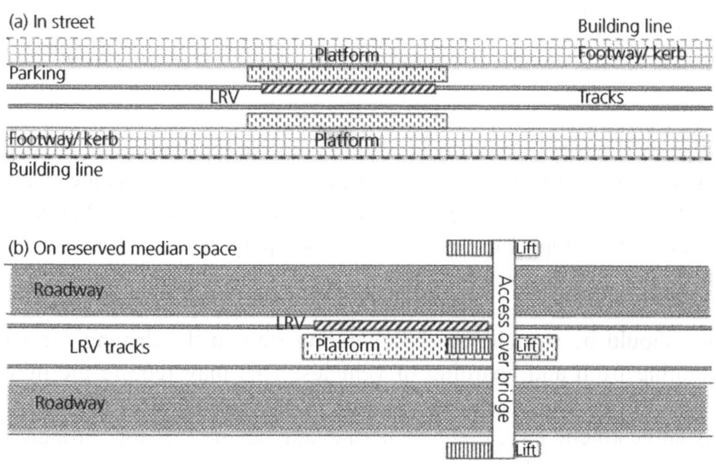

Source: The author

the station into the street scape using existing footways or sidewalks is better, where possible. The formal demarcation between the platform and the functional space available from the footway would allow passenger build ups, without overcrowding. This will also be the most convenient and safest way to access the platform.

By definition, about 50% of the passengers will originate from or migrate to the opposite side of the road from which they arrive or depart. They will need to cross the tracks to reach their destination after alighting, or to board. Historically, heavy railways have made passenger crossings of tracks grade separated, by having bridges or subways, as at Virginia on the San Jose system (Figure 4.27). Here, however, by using an island platform a spacious waiting environment is provided. Being in the median of a busy freeway means that an overbridge is needed (Figure 4.28(b)) for a grade separated and safe access to the station from its catchment area. This will also require lifts to enable passengers with mobility problems to ride the LRT.

Unless there is some particular local problem, light rail systems normally allow passengers to cross the tracks at grade. The most obvious reason for this is that it is preferred by passengers and is easiest for those in wheelchairs or with baby carriages. As LRVs stop at most, if not all, stations with powerful emergency brakes, there are unlikely to be many safety issues, especially if a speed limit applies to the vehicle approach to stops.

Passengers using a light rail service are more likely to be involved in a safety incident with other road vehicles while gaining access to the station. Having surface crossings of tracks also reduces the size of platform needed, since space is not taken up with footsteps/ramps up to a bridge or down to a subway. Finally, considerable urban security research shows that most people do not like subways, due to the claustrophobic effects or perceived threat of assault or theft. Pedestrian subways or bridges are also difficult to make accessible to passengers in wheelchairs and so lifts are needed to provide full mobility accessibility.

About 80% of the adult population can make a step of 250 mm (above the track) to reach the platforms and 90% 200 mm steps. Station design should however guide the passengers to cross at dedicated crossing points, normally at one or both ends of the platform. Depending on location, the tracks may be paved, or open and ballasted. In either case designs should discourage passengers from walking along the tracks away from designated crossing points. This can usually be achieved by having a fence between the tracks at stations, where the station tracks are not shared with other road vehicles.

Should interchange with buses be envisaged, where possible this should be cross platform in the peak flow direction, ideally with simultaneous timetabling. Since interchange is always perceived as a penalty in all market research and, hence, generalised cost calculations. Making an interchange easy for passengers without imposing difficulties on the operating staff is a wise consideration for the design engineer. It is also very user friendly.

Finally, platforms should be straight, so the operating staff in the driving cab can fully supervise the safe alighting from and boarding of vehicles. This may require the minor redesign of curved approach tracks and the possible tapering of the platform ends, to accommodate the LRV swept path, or dynamic kinetic envelope on entering and leaving the stop. Straight platforms with tapered ends are, however, safer than curved platforms where it is difficult to maintain safe gaps for passengers, including those in wheelchairs, to board and alight.

4.6. Depot

Light rail vehicles must be stabled overnight, valeted (Figures 4.29 and 4.30), maintained and repaired. This requires a depot (Figure 4.29) with suitable facilities. The minimum requirements, to ensure a safe and reliable service, include a covered building for vehicles to be dismantled for maintenance or repair. A wheel lathe will allow the running gear to have a profile for smooth and safe running. Staff, both maintenance and operating, need facilities and space for training or briefing, as well as hygiene. Vehicles coming back to the depot at night will normally be washed and internally cleaned.

Figure 4.29 Possible depot layout

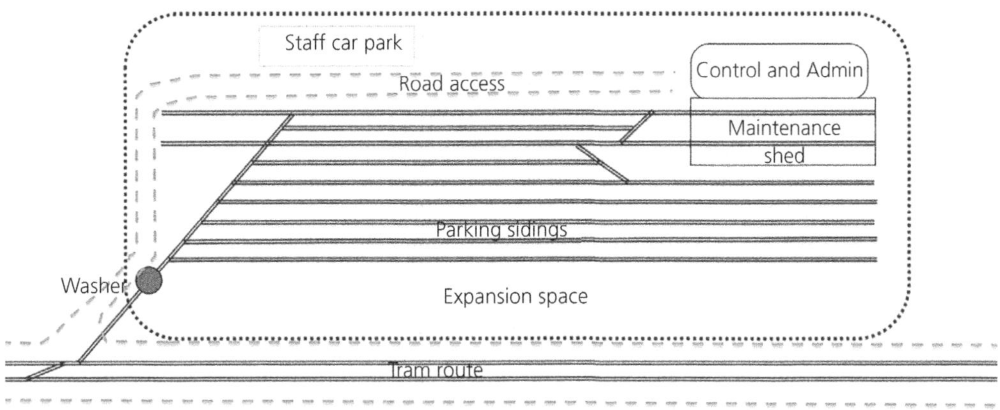

Source: The author

Figure 4.30 Depot rail vehicle washer

Source: Yarra Trams Melbourne

The light rail system designer is under a very severe constraint compared to a bus service planner. Bus depots can almost be located anywhere in an urban area and still be able to service several routes. The tram or light rail depot must be physically connected by track to the passenger service operating lines. Operating between the line and the depot is a dead cost and generates no revenue. Building a connecting line also increases the capital cost. Depot access also takes time out of the driving shift.

In designing the light rail line, the engineer needs to be able to find suitable low-value sites, where a depot can be located close to or preferably next to the line. At the same time a depot being semi industrial may not be in residential areas, where it may be seen as a poor neighbour. Ideally, the site should be flat or made flat and have access from two ends if possible. The site should allow for future expansion to stable and maintain a larger fleet. Staff facilities may also need adjustment. If extra space is not possible, then the initial depot can become the central maintenance facility and new stabling depots added to new lines as built.

The exact facilities needed will depend on the nature of the vehicle fleet. A new system will surely opt for low-floor trams, so that low platform stops can be built with excellent accessibility. Low floor vehicles often have most of their equipment on the roof, where the pantograph will be located. Therefore, the depot will need gantries to enable staff to work safely at height. Any under vehicle work can be undertaken by hoisting it off the floor (Figure 4.32), rather than having pits (Figure 4.31).

There should be at least one depot line with OHL where repaired trams can be run to allow for testing before being released into passenger service. The inside of the depot will not have OHL, so LRVs will be shunted in by a small tractor. If a wheel lathe is not provided, wheels developing flats must be removed and sent to a specialist company. With more LRVs having independent wheels, the need to maintain the same radius on both sides is reduced. A solution to this is having a spare set of wheels, that can be used to replace 'flat' wheels, so that the LRV can be returned quickly to service.

Figure 4.31 Blackpool depot maintenance pits

Source: The author

Figure 4.32 Tram on jacks, Edinburgh

Source: Edinburgh Trams

In some cities, brown field or derelict industrial sites may be available, where a new depot could be used as a fulcrum to help stimulate urban regeneration. Indeed, a deal might be struck with a property developer, with the ground level of a site used for the depot and the air rights for commercial and other developments. This would have the benefit of excellent access to a new light rail line. Some European cities are adopting this approach, with the space above tram depots being used to provide residential accommodation. Trams emit no air pollution, are quiet in operation and can therefore be an acceptable neighbour, helping to increase the urban housing stock.

Where such sites are not obviously available next to the preferred first line, the depot may end up dictating part of the alignment of the initial system. It would make considerable economic sense to divert the line to be closer to the depot. There is a trade-off between a reduction in (depot) capital costs and any dead mileage accessing the depot and a different pattern of patronage. This may not be the most preferred route but, later, the preferred route can be opened linked by way of the first route to the maintenance depot.

For new systems, the depot site should be chosen to allow for subsequent network expansion, since this will bring economies of scale in terms of technical staff, stores and maintenance equipment. Space to lay further over-night storage tracks is a low-cost investment but later may have a very high value, by not needing to build another depot on a new site.

Once the first line has been built and opened, subsequent network extensions might include links that were omitted when the first line was built, in order to provide a depot site economically. This will almost certainly include some shared trackage and an increase in service frequency on a trunk section of line, which will increase patronage and revenue.

Often new systems are over indulged in terms of depot facilities. Clearly there are basic daily, weekly, monthly and annual maintenance activities. There may also be a contract with the vehicle supplier to undertake heavy maintenance elsewhere. Daily maintenance will include washing the outside of the vehicles and cleaning the inside. A drive through washing machine is normally used, which may be outside or in. An internal washer may be preferred for environmental reasons (Figure 4.30). There are many vehicle washers on the market, of which some are proprietary makes. The location of this washing facility will be determined by the operational procedure in the depot.

Many systems wash their vehicles at the end of the day's schedule as the vehicle enters the depot. Similarly, most systems clean the inside of vehicles with over-night valeting teams, so that vehicles are pristine for passenger service first thing in the morning. Other systems wash the LRVs as they leave for service.

For those systems which have vehicle wheel sand applicators, checking and refilling sand boxes will have to be located normally on the inbound tracks prior to overnight parking. Again, proprietary equipment is available to keep sand dry in the depot and deliver by pipeline under air pressure directly into the LRV sand box.

During the operating day, drivers will report minor faults and problems with their vehicles. Modern vehicles have automatic equipment monitoring and remote reporting to central control. This helps the depot maintenance manager assess the need to examine specific vehicles on return and prepare a maintenance rota. Obviously, vehicles that fail and have been towed back to the depot will already be allocated workshop space for repair or rectification.

Minor problems may be examined by a fitter or mechanic on the inbound track and either repaired there, or diagnosed and remedial action determined to be undertaken in the workshop, on a rota basis. Part of this may be automated, since many LRVs now have onboard diagnostic equipment. These send condition messages, that will be logged and used to determine the optimum maintenance schedule.

Maintenance and repair functions all need space and facilities to be performed safely and efficiently. Workshop space under cover and heated/cooled for repair and maintenance will both make for a better workforce and improve the quality and reliability of maintenance, and enhanced fleet operations and performance. The key dimension for this workshop space is the size of the vehicles to be maintained and whether they can be easily split into smaller parts. Most LRVs today are at least 30 m long, so allowing for space at either end, means a depot length of a minimum of 40 m. The second dimension is workshop width. This will be determined by fleet size and, therefore, the planned spare vehicle and maintenance programme. Normally, however, such LRVs are in several sections that can be split, which will reduce the length of the dedicated workshop section or enable more than one vehicle to be worked on each track at the same time. This is especially so when working under the vehicles.

Workshop pits were the original way to undertake repairs and maintenance beneath LRVs (Figure 4.31). Pits however have major drawbacks, including safety and inflexibility. They can also flood and will need regular and expensive cleaning to avoid the build-up of oil and other dangerous debris. If pits are deep enough to walk along, then moveable staging will be needed to be able to work comfortably and safely under LRVs. If on the other hand the pits are designed for a comfortable working height, walking along them may not be easy with a stationary LRV on the track. Most systems do not allow staff to be in pits under LRVs that are moving.

Another depot approach is a pit-less workshop, with vehicle jacks (Figure 4.32). These allow vehicles to be lifted to a safe working height off the tracks and can be adjusted depending on the task to be undertaken. This is particularly valuable when bogies need to be removed or replaced. A flat depot floor with vehicle lifting jacks will require less shunting since the jacks can be taken to the vehicle, rather than the vehicle being moved over a pit. With a flat depot area and the use of hover-pads, vehicles can also be moved anywhere in the depot and then jacked up to working height at the most convenient workstation. A no pit workshop also reduces the danger of people falling in and being injured, or the need to install safety railings to prevent this, reducing the utility of the workshop and its pits.

There are also soft requirements for a depot. The workforce will need to change clothing at the end of a day's work, and washrooms, toilets and showers are a necessary provision. A canteen may also provide meals for all staff, including vehicle drivers. There will be a control centre where operating duties are allocated, services managed in real time, and signalling and electrical power controlled. There will be a store both for vehicle spare parts, as well as staff protection equipment and clothing. Finally, it is usual for the system general manager and other senior staff to be located at the main depot, together with staff training facilities, both for maintenance personnel and vehicle operators.

Ironically, most light rail depots will need a staff car park, especially as some crews will start early in the morning and others finish late at night, needing transport to or from and their homes. Even if the operator has a policy to employ staff who can use the light rail service to reach the depot, or provide a minibus to ferry staff, there will always be those who cannot make use of that facility and must drive. Visitors to the depot, as well as suppliers of parts and so on, may also need parking space, with provision for any road based maintenance or emergency vehicles.

4.7. The vehicles

4.7.1 Brief history

The first trams/streetcars were larger versions of existing horse drawn road vehicles, with wooden bodies and wheels bolted to the floor frame. Such a vehicle would weigh less that 2 t (4400 lbs) carrying about 30 passengers, which could be pulled on the level by a single horse (Figure 4.33).

Figure 4.33 Horse drawn tram, Toronto 1890

Source: Picryl

When electric traction was adopted at the end of the nineteenth century, the extra weight of the motors meant a change in construction techniques. While the body remained wooden, it was mounted on an undercarriage of iron or steel needed to support the motors and wheel sets. The wooden body was bolted to the underframe.

This approach lasted until the 1920s when car builders looked at automotive practice, where new models were launched every few years, allowing progressively more advanced all metal body structures. Firstly, in the UK the private tramways of London took a 'back to the drawing board' approach and adopted a single integral metal body frame and cladding, to give strength and reduce weight, allowing a more luxurious passenger vehicle. One hundred of these 'Feltham' trams (Figure 4.34) were built in 1931, which provided a new benchmark for light rail vehicles in Europe. These continued to operate until the Leeds system closed in 1959, giving 30 years of service without the need for refurbishment.

In the USA in 1929 a committee of private streetcar operators (Presidents' Conference Committee (PCC)) agreed that something better than the nineteenth century vehicle was needed. This resulted, in 1934, in a very radical vehicle redesign. Like the UK an all metal body was adopted. The bogies were rethought and motors were frame mounted driving wheels set by way of drive shafts (Figure 4.36). Rubber sandwiched (resilient) wheels (Figure 4.35) replaced solid, giving a smoother ride. The electrical control system went from a ten-stage resistor to 100, with very smooth and fast acceleration, following a programme of ergonomic experiments. These determined safe and comfortable levels for especially standing passengers. This PCC car had an outstanding performance (Table 4.1). The first system to adopt PCCs was Brooklyn in 1936 (New York) (Figure 4.37).

The PCC set up the Transit Research Corporation, which licenced a number of car builders in North America to build to the PCC design. Nearly 5000 vehicles were thus produced, keeping

Figure 4.34 London 'Feltham' tram 1931

Figure 4.35 PCC resilient wheel rubber sandwich

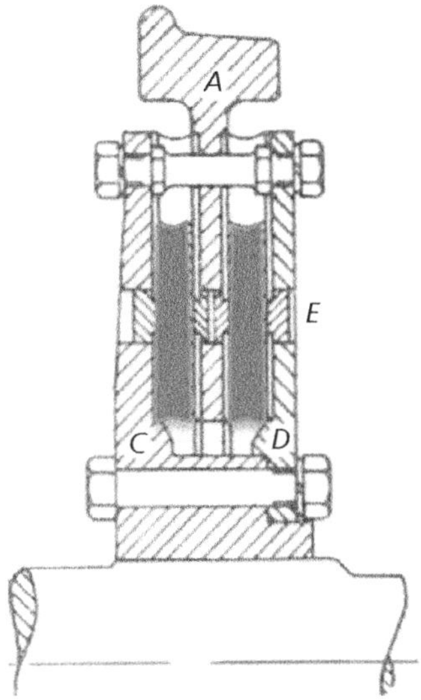

Figure 4.36 PCC bogie

many streetcar systems running well into the 1970s. PCC cars were still running in the 2000s, a tribute to the robustness of the design. After the second world war, the PCC design was further licenced to European car builders in Belgium, Czechoslovakia, Italy and Spain, where some 60 000 variants were built, some still operating.

Post 1945 many cities were devasted, and tramways were often the first urban transport to be brought back to operation on a make do and mend basis. By the mid 1950s these trams were

Table 4.1 PCC streetcar performance

Criteria	Data
Passenger capacity	50 seated and >50 standing
Empty weight	18t (40 000 lbs)
Length/width	15.4m (50,5ft) / 2.75m (9ft)
Power to weight ratio	9.0kW/t
Acceleration	2.1 m/s/s (7 ft/s/s)
Service braking	1.8m/s/s (6 ft/s/s)
Max speed	65km/h (40mph)

Figure 4.37 Brooklyn NY PCC Car 1936

wearing out and new vehicles were needed, and the quest was on to lower the building and operating costs. Electric motors are expensive and so minimising the number needed was a goal. The Feltham tram (Figure 4.34) used 'monomotor' bogies' (Lesley, 2025) of the 'maximum traction' concept, where only one axle is powered but the bogie king pin is off centre, so that the driven axle carries about 75% of the weight and the second axle with smaller wheels only 25%. These have been in use since the late nineteenth century.

Most trams have wheels attached to rigid axles. This needs maintenance to keep both wheels to the same diameter and profile, otherwise there is unsteady riding on straight tracks and torsional windup of axles, although much worse on the small radius curves usual on tramways.

The Düwag company in Düsseldorf in the mid-1950s took up work, started in Liverpool in the 1920s (Figure 4.38), on an equal-wheel bogie with the motor at right angles driving both axles through hypoid gearboxes (Figure 4.39). These monomotor bogies increased the need to keep all wheels at the same diameter and profile, since not only could torsional windup of axles occur

Figure 4.38 Liverpool monomotor bogies 1929

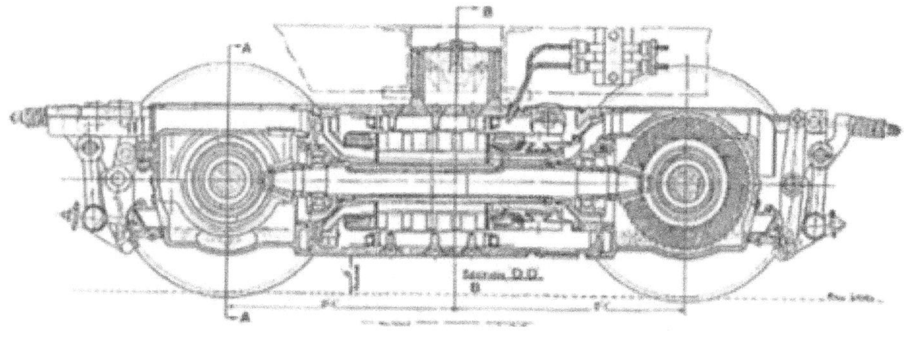

Figure 4.39 Düwag monomotor bogie

but also in the motor shaft driving both axles. The cost of motor replacement in such bogies is considerable.

Bogie weight directly impacts the track, so there is a need to reduce the un-sprung mass on wheels. Solid axle wheel sets have a mass of some 2 t (4400 lbs). The PCC approach used rubber cushioned wheels (Figure 4.35), reducing the un-sprung mass to the weight of the wheel tyres. This has been continued in the Düwag and Trampower monomotor bogies (Figure 4.40), using different designs of sprung wheels.

Trams and LRVs can be single or double ended. Most PCC cars were single ended, meaning that terminals need turning loops or a city centre block run around. Other historic systems used double ended cars which simplified terminal arrangements. Double ended cars will have doors on both sides. This makes stop location and design easier.

Figure 4.40 Trampower monomotor bogie

4.7.2 Contemporary LRVs

While minimising the un-sprung mass, minimising the bogie weight is also important, as this contributes to the total vehicle weight and therefore the power needed for acceleration. The Liverpool monomotor bogie weighed about 5t (11 000 lbs). The Düwag bogie more than 6t (13 200 lbs), while the Trampower bogie only 2.5t (5500 lbs). These are important contributors to the total vehicle weight.

Each configuration has a different seat and standing space (usually at 4/m^2) and, hence, capacity.

Added to the innovation of monomotor bogies was the development of articulated trams, with two powered (outer) bogies and a centre bogie only supporting the articulation unit. Such vehicles only needed two motors and allowed operators to improve the productivity of staff. Only the driver is needed, with the vehicle carrying nearly double the number of passengers compared to rigid trams, even when run as a two-car 'train' (consist).

The labour cost savings were much greater than the increased cost of bogie maintenance mentioned above, which was the main reason Düwag articulated trams on monomotor bogies became the market leader in the period 1960–1990.

Table 4.2 Comparison of different car weights

Car Type	Length	Axles	Empty Weight	Weight/m
Sheffield Düwag	34.75 m (225 ft)	8	52 t(114,000lb)	1.5
Manchester Bombardier	29.28 m (97 ft)	6	39.7 t (87 340 lbs)	1.4
Dublin Alstom	43 m (142f t)	8	53 t(116 600 lbs)	1.2
Trampower	29 m (95 ft)	5	22 t(48 400 lbs)	0.8
Single deck bus	12 m (39.4 ft)	2	10 t(22 000 lbs)	0.8

Source: The author

Figure 4.41 Düwag two section trams on two monomotor bogies

The articulated tram transformed the economics of light rail services and this logic allowed further development by the addition of a third section between the two driven outer sections, improving the productivity of staff.

By the 1980s, a new social demand became an objective for the development of light rail systems, to provide access for passengers with disabilities. This is not easy with trams where the inside floor is 900 mm (3 ft) above the level of the road, and two or three steps are needed to board or alight. Two approaches have been adopted to address this need. In Germany, Hannover and Stuttgart have adopted high platforms (*c.* 900 mm/3 ft high) at tram stops, as has Manchester in the UK, Calgary and Edmonton in Canada, and San Diego in the USA. For many cities, fitting high platforms in narrow streets is not practical or acceptable and has safety implications when platforms are crowded, with a danger of passengers falling off.

The alternative has been the development of low-floor trams and the adoption of low platforms (300 mm/1 ft high) at stops. Initially, existing articulated trams were fitted with low-floor centre sections (Figure 4.42).

The first attempt at a low-floor design was in Geneva, where Düwag-Vevey designed small wheels for the articulation bogie and conventional large wheels for the outer powered bogies. This arrangement has about 70% low floor. Alstom developed a similar vehicle for Grenoble. The LRVs used in Sheffield need all wheels motored to climb 10% gradients, so compromise was adopted in that instance with a short low-floor section in the two end sections.

The first truly low-floor tram was developed in Vienna (Figure 4.43). This required the design of a new articulation unit incorporating vertical side-mounted motors powering individual wheels through special gearboxes.

These trams were expensive to build and maintain which led to the adoption of a lower cost options, namely fixed truck bodies with wheel arches for seats and a low-floor aisle between (Figure 4.48). In practice, these designs have created many problems, just like the fixed truck vehicles over 100 years earlier, only resolved by the adoption of bogies.

Figure 4.42 Düwag high floor tram with low-floor centre section in Gotha

Source: Urban Transport Magazine

Figure 4.43 Vienna low-floor tram operating in Romania

To achieve a '100%' low floor, numerous designs were produced by several manufacturers. These needed different traction concepts, with body-mounted motors using cardan shafts linked to bogies. Wheel hub motors have also been tried. Often these complex arrangements resulted in poor reliability and high maintenance costs.

Alstom tried to standardise low-floor designs with its 'Citidis' family. This approach enables various vehicle lengths and capacities, using between two and seven sections. This standard approach is customised by difference front end styling, interior layouts and liveries.

Siemens followed with its 'Combino' range. The first operator was Potsdam in Germany. Unfortunately, after about 100 000 km of operation, fatigue problems required a rebuild. By that time, some 400 units were in operation, all of which needed retro-repair. The body structural problems aside, fixed four-wheel truck have significant operation and maintenance problems discussed fully elsewhere (Lesley, 2019). These include low speeds around curves, rail and wheel grinding, and the development of rail-head corrugations from unstable 'hunting'. Figure 4.44 shows the variant operating in Almada, where there is a truck under each body section. Costs were further reduced by only having a truck under every other section, as seen in Edinburgh (Figure 4.45).

Since then, designers have been trying to find a way to incorporate bogies in a mostly low-floor tram configuration. First off was Trampower in 1989 with a 70% low-floor design, incorporating powered lightweight (2200 kg/ 4840 lbs) bogies (Figure 4.40) at the outer ends and a steered single-axle centre articulation unit.

Subsequently, car builders have developed other designs, usually requiring seat arches to accommodate the wheels of rotating bogies (Figure 4.48).

Vehicle stability is important when running on public roads. In the UK, single deck trams must be able to withstand a tilt of 35° and double deck 28 °, when loaded for the worst case. This means the designers need to ensure that the centre of gravity vehicles is low. Figure 4.49 shows an example for a single deck car, where the centre of gravity must be no higher than 1200 mm above the track to be stable on a 35° slope. Figure 4.49 also shows a real double deck tram tilt test in 1933 to 28°.

Figure 4.44 Siemens fixed truck '100%' low-floor tram Almada

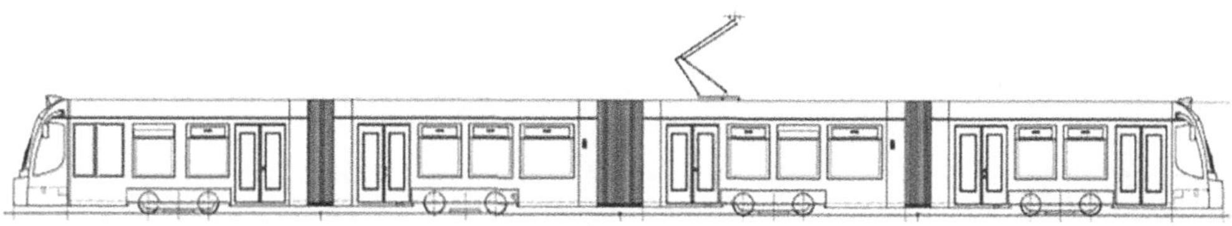

Figure 4.45 Edinburgh CAF 7 section tram on four trucks

Figure 4.46 Testing the Trampower low-floor LRV in Blackpool

The author with Paul Jackson RIP.

Source: The author

Figure 4.47 Examples of part low-floor trams with bogies

Kinki Sharyo 60%

Bombardier 65%

Düwag/Siemens 40%

Trampower Mk 2 70%

Figure 4.48 Example of wheel arches to accommodate rotating bogies

Source: The author, Skoda tram at Innotrans, Berlin, September 2024

Figure 4.49 Tilt test of single and double deck LRVs

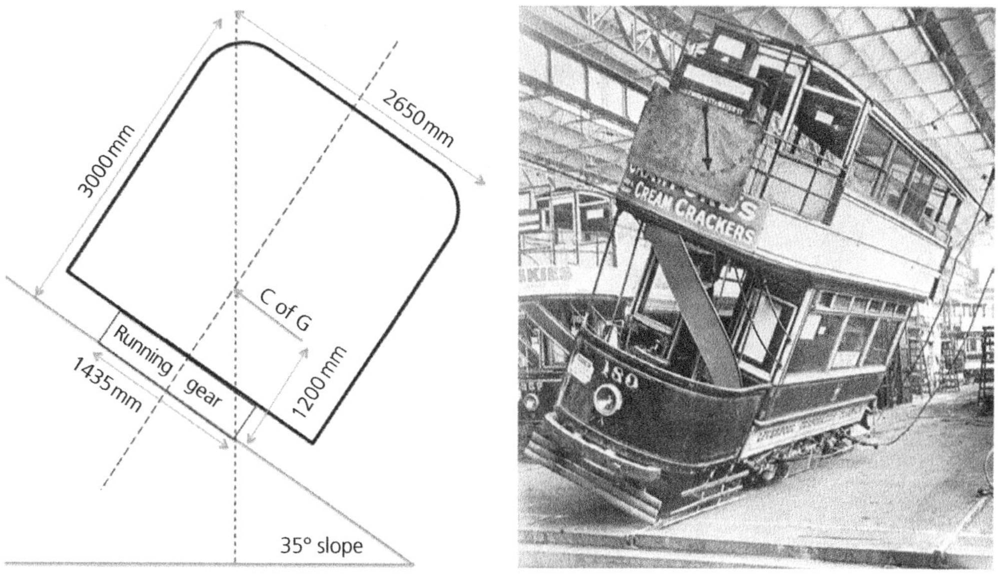

Source: The author and Merseyside PTE

4.7.3 The ideal LRV

The discussion above demonstrates that many contemporary trams/LRVs have problems both in terms of the need for continuous vehicle maintenance but also from wear or damage to tracks. Part of the reason is that LRV demand is a small market, so there are few car builders, who find it more economic to downscale heavy rail vehicles produced in quantity, rather than design bespoke for the special needs of urban tramways. These include small radius curves (< 20 m), steep gradients (<14%), frequent stops and the need for fast acceleration and braking to give high operating speeds to compete with private car travel. Lighter vehicles are needed to minimise track wear and the use of energy.

When developing a new LRT system, a search of existing designs and secondhand vehicles should be made. This may reveal an option to suit the system or can be modified cost effectively. This can help to avoid the current proliferation of designs. Hence Manchester, in 1989, adopted the Bombardier (now Alstom) design used in Cologne. Clearly there was no vehicle on the horizon to match the PCC car of the 1930s.

When designing a new tram or LRVs, there are several areas to consider.

4.7.3.1 Ergonomics

There has been much work on the ergonomic requirements for passengers and staff, pioneered for the PCC project in the USA. Much of this is still relevant, including step heights, grab rails, seat design and spacing. Confirming this by producing mockups and getting potential passengers and drivers to assess should be a key part of supplying vehicles. This should include configurations for the benefit of passengers with disabilities.

Ergonomics show that standing passengers can comfortably cope with 1.8 m/s^2 (6 ft/s^2) acceleration. Compared to much slower rates in contemporary trams, the City Class was tested to 2.2 m/s^2 (9 ft/s^2). Figure 4.50 shows the braking graph of a City Class tram at 1.5 m/s^2 and a jerk rate of

Figure 4.50 Braking graph of City Class tram

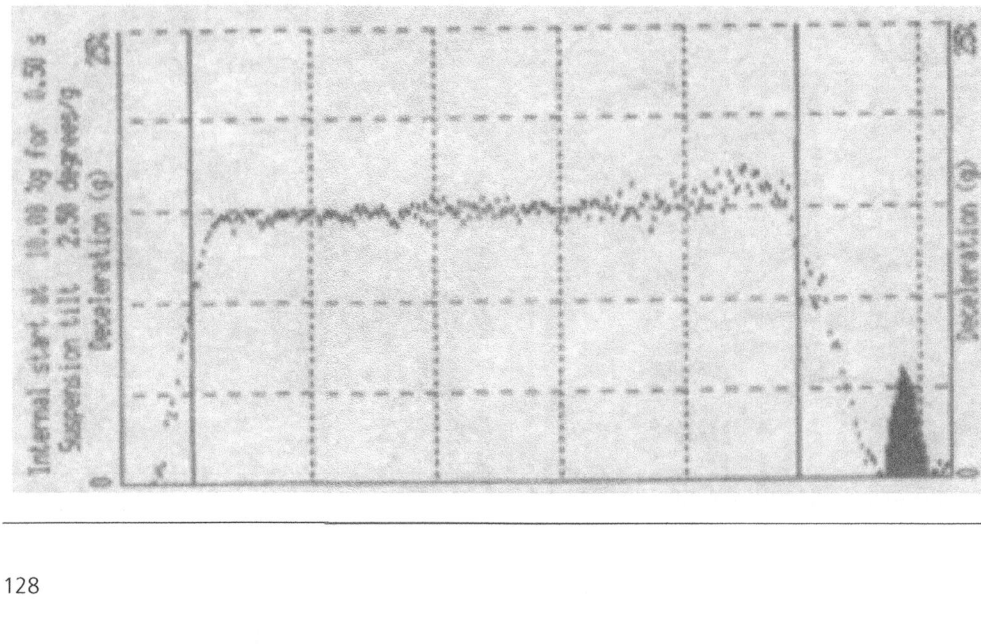

under 1.0 m/s^3. The final triangle is suspension reaction. With solid state controls, smooth acceleration and braking, low jerk rates of under 1.0 m/s^3 (3 ft/s^3) are possible, giving standing passengers time to adjust their grip.

4.7.3.2 Economics

There have been radical changes in the traction and control of LRVs. Electro-mechanical controllers with direct current (dc) motors have given way to solid state electronics and alternating current (ac) motors. Often this leads to higher costs, as manufacturers seek to shoehorn equipment into small spaces. Such bespoke equipment then becomes single source and costly to maintain.

Trams and LRVs compare badly with the cost of new buses, which on a like for like basis are about 15% of the price. One of the reasons for this is that there is a large market for new buses, as Table 4.3 shows world production ten times bigger than all types of rail passenger vehicles.

A second reason for the lower cost of buses is that they share many costly parts – for example, engines, gearboxes, transmissions, brakes, suspensions and so on with road trucks, for which there is an even bigger world market. Buses, uniquely, need a special body which is a low value item. For trams, a useful break down of costs is: body 33%, electrical and control 33%, and running gear 33%.

The market size for LRVs/trams is too small (Table 4.3) to make mass production a way to reduce costs. The nearest to that was between 1948 and 1989 when Tatra in Czechoslovakia supplied the whole Soviet Comecon block with an average output of 1200 vehicles per year. Today there is no single (closed) market as large. An alternative is to use mass produced parts from other industries and fit them to create an economic vehicle. This was the approach adopted for the City Class tram.

Components off the shelf (COTS) have another benefit in reducing the cost of spare parts available from multiple suppliers. The original equipment manufacturers (OEM) of COTS are into continuous product development, so that future technically compatible replacement parts are likely to be better.

Ironically all tramways are powered with dc, but the new alternative current (ac) control philosophy could equally be powered by single phase ac.

Solid state control powering sealed ac motors offers the operator lower costs from more efficient use of current and less maintenance. For passengers, stepless, jerk-free acceleration and braking to a standstill is more comfortable and well within the limits of what is ergonomically safe. Alternating current drives are quieter than dc motors and stay quiet, since there is no commutator or brushes to wear. Such control systems will also optimise the supply of current to provide motors

Table 4.3 UN Data of world production 2019

Vehicle type	Numbers made
Rail passenger (all types incl. trams)	20 000
Buses	200 000
Road trucks	20 000 000

with just enough power to achieve the desired acceleration, speed or braking irrespective of passenger load. This makes running timetable adherence easier. Monitoring the condition of motors allows for slip/slide protection to be built in, sparing rails and wheels from mechanical damage, including wheel burns.

Adding micro-processor control of the operation of LRVs to this can greatly simplify the design and construction, with the adoption of distributed control for the operation of doors, lights and so on. Effectively, this replaces a pair of dedicated wires for each function, with four wires for the whole vehicle, two providing power and two using distributed logic control with individual identification of each function. Such control can also monitor and report failing or failed equipment. This diagnostic capability should lead to preventative maintenance, to keep vehicles running for the maximum time, thus reducing the need for spares, over and above those needed for routine maintenance. Replacing 'hammer and spanner' functionality with micro-processors improves the reliability and reduces the operational and maintenance costs.

Lighter vehicles use less power than heavy ones. Unfortunately, trams and LRVs have become heavier (Table 4.2). The majority of vehicles weigh about 1.5 t per metre length (1020 lb/ft). In contrast, the Trampower vehicle at 0.8 t/m will use about half the power of heavier trams. During the operating life of the vehicle, this power saving will exceed the cost of a new tram. This emphasises the importance of determining the whole life cost of a system, not just its upfront capital investment cost (CAPEX).

The final part of economics is staff productivity. All tramways will need drivers until fully automatic street running tramways are acceptable. The protocols to make automatic operation possible have been set out (Bell, 1975). Historically, rail vehicles have got larger, needing fewer staff to provide better productivity, underwriting the pay needed to attract and retain suitable staff. Added to this, the skill needed to drive a tram/LRV is not easily transferred to another industry, so that turnover of staff is much lower than many transport operations. The productivity of staff can be improved by both using larger capacity vehicles and raising the operating speed, so that fewer vehicles and drivers are needed for a given service frequency.

Should passenger demand grow more than expected, then extra capacity can be provided by buying more trams, by stretching existing trams (Figure 4.51), or by making the schedule more productive with a faster operating speed.

Figure 4.51 Stretching the City Class from 29 m to 38 m (200 to 300 passengers)

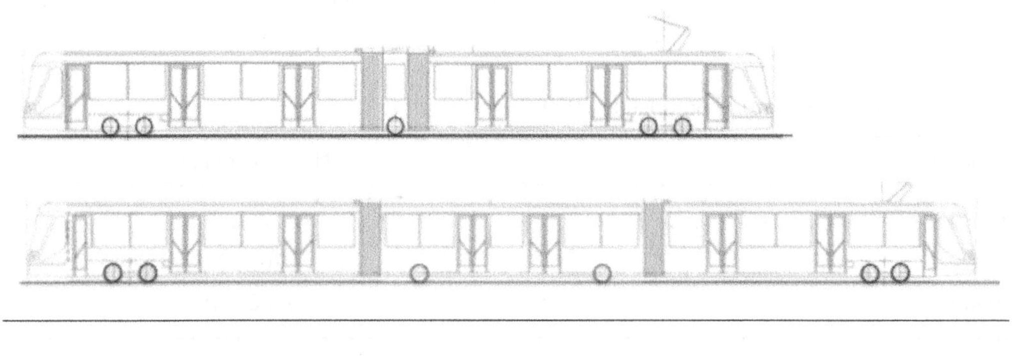

4.7.3.3 Track wear and so on

The vehicle design team will not only need to produce vehicles that perform and are passenger and staff friendly but are also easy to maintain and track friendly. Rigid four-wheel trucks, while making it easy to accommodate low floors, produce poor riding, needing slow speeds into curves with rail punishment due to wear and fatigue damage that needs restorative grinding, shortening rail life. This was the lesson learned more than 100 years ago leading to the adoption of bogies, swivelling four-wheel subframes, able to ride into curves with less wear due to shorter wheelbases. Bogies also give smoother riding for passengers, since four-wheel fixed trucks lead to 'hunting', which is uncomfortable for passengers and causes wearing of tracks. These matters are discussed more fully in 'Embedded rail tracks' (Lesley, 2025).

There are car makers claiming to build 100% low-floor trams. This is misleading, since floors have to accommodate wheel arches, leaving only a narrow aisle along the vehicle, usually too narrow for a wheelchair or baby buggy to pass. Bogies give smoother riding and less track wear. This means that there can be less of the vehicle with a low floor and so fewer wheel arches. Experience shows that many passengers are willing to climb a couple of internal steps to reach a higher floor section, needed over bogies, leaving the low-floor step-free section for the less mobile and passengers with baby carriages.

4.7.3.4 Environmental

The environmental impacts of transport systems are large, from air pollution to health-threatening toxic emissions and noise. Even 'electric' buses create more than WHO levels of micro-particle pollution from abrading tyres on tarmac. The quietest bus passing noise is 70 dBA just from the tyres. The WHO noise standard for daytime cities is 55 dBA. Most buses are diesel powered and so produce tailpipe NOx and micro-particles and by using a liquid fuel emit CO_2, adding to global warming. Alternative bus designs include diesel/electric or straight battery with or without recharging on the route.

Three sources of tram noise can be a great nuisance. One is due to vehicle design and the other to track constraints. Wheel squeal on curves is caused by wheels solidly fixed to the axle. One wheel will rotate at the right speed for the track, the other micro slips and slides across the rail head, resonating the rail (Lesley 2009). Cures like rail greasing reduces it but also can reduce braking efficiency. Wet rails ring less, as water is a benign lubricant. One cure for this is independently rotating wheels, which many contemporary trams now use and was a design constraint of the City Class from the start.

The second noise nuisance occurs at rail crossings using heavy rail designs. The cause is the wheel dropping into the crossing groove. The cure has been available for more than 100 years, namely 'silent crossings', where the groove bottom is raised so that wheels are flange supported over the crossing gap. There is a longer discussion on this in 'Embedded rail tracks' (Lesley, 2025).

The third is 'roaring rails', due to rail head corrugations (Figure 4.22). This is usually a fatigue failure (Lesley, 2009). The principal cause is the uneven running of wheels (hunting), with micro slipping, creating work hardened sections. This can be mitigated by rail head grinding but shortens rail life. Avoiding the creation of corrugations is possible with independent wheels. Test running the City Class tram over sinuous track and then examining the bogies (Figure 4.40) showed the wheel tyres had a 10 mm (0.4 in) wide wear strip in the centre of the tread and that the wheel flanges were rusted. This indicated that the bogie rides very smoothly along the centre of the track and that there is no flange/rail side contact on curves. This means that both the wheels and rails will last longer.

In comparison to other road vehicles, trams produce no air pollution in the street. The ferrous particles from tyres and rail interaction are heavy and rust benignly in rail grooves. With renewable power generation to supply trams and LRVs, the systems are carbon neutral, and by attracting trips from cars, carbon negative. Also, trams have noise levels that usually meet WHO standards, being about 60 dB(A) maximum.. By all normal measures, trams and LRVs are good friends of the environment.

4.7.3.5 Single or double ended and single or double decked

Most trams/streetcars/LRVs are double ended and double sided, giving the most flexibility in operation. This is especially so at terminal. A double-ended vehicle only needs the driver to walk to the other end of the car and then proceed in the reverse direction. A double-ended vehicle has doors on both sides, so that stops can be on either side. This maximises the flexibility of stop locations and layouts. Stops can either be part of the footway/sidewalk, in which case about 50% of the passengers will need to cross 100% of the road. An island platform in the centre of the highway will need protected access for 100% of passengers crossing 50% of the road.

Single-ended vehicles were popularised by the PCC in the USA. These only have doors on one side of the car and a driving cab at the leading end. This increases the space for seating and standing passengers. It does, however, need a reversing facility at terminals. These are mostly loops. One exception to this was Rotherham which had turning triangles at terminals. That tramway closed in 1949.

After 1955, the new articulated vehicles from Düwag were mostly built single ended and single sided to maximise passenger capacity. These needed the operating tramways to construct turning loops at terminals. The trams were designed to pass around 15 m radius curves.

Britain and places influenced by Britain also used double deck vehicles. This has the advantage of nearly doubling the passenger capacity for a given vehicle length. With the need for stairs, two on a double ended vehicle, boarding and alighting is not as fast as multi-door single deck vehicles. There are no regular double deck tram services in Britain, although Blackpool runs some of their historic trams as a tourist attraction. Outside Britain, Hong Kong (Figure 4.52) continues to use double deck trams.

Alexandria, in Egypt, also operates double deck cars. As can be seen in Figure 4.53, these are multi-door vehicles, for rapid loading/unloading and capable of hauling single deck trailer cars. These however are being withdrawn, and replaced by new single deck vehicles.

The choice of single/double ended cars will depend as much on the possible geometry of the road network and the ability to accommodate reversing terminals, where turning loops can be fitted. In some tramways there can be both loops (Figure 4.54) and dead-end reversals.

4.7.3.6 Conclusions

The specification by clients for new trams/LRVs can either be performance based, or in fine technical detail and effectively a whole vehicle design. A performance-based specification will include

Figure 4.52 Two Hong Kong trams

Source: https://hongkongfreetours.com/hk-tramway-and-tramoramic-tour/

Figure 4.53 Double deck tram, with single deck trailer cars in Alexandria, Egypt

Source: https://www.urban-transport-magazine.com/en/tram-modernisation-in-alexandria-egypt/

availability and reliability, passenger capacity, seat/stand split, acceleration and braking rates, maximum speed, weight and energy consumption. This is important because, over the lifetime of the vehicle, the cost of power will exceed the purchase price. An example of the result of an earlier market research exercise undertaken by Trampower Ltd. is shown in Table 4.4.

Figure 4.54 Turning loop on Oslo Tramway

Source: https://www.google.com/maps/search/oslo+tramway/@59.9475911,10.715093,104m/data=!3m1!1e3?entry=ttuandg_ep=EgoyMDI0MTAyMS4xIKXMDSoASAFQAw%3D%3D

Table 4.4 Comparison of specifications

Requirement	Transit authorities	City Class tram
No. of passengers	200	200
Seats	50	80
Tram length	29m	29m
Weight empty	<50t	22t
Low floor	>50% of length	75% of length
Low floor height	380mm	300mm
Minimum radius	25m	15m
Maximum gradient	6%	10%
Maximum speed	80kph	90kph
Acceleration	1.1 m/s/s	1.5m/s/s
Power consumption	c. 4kWh/km	c. 1kWh/km

In procuring a new vehicle fleet, building a mockup is a good way to test both markets acceptability by potential passengers and also the ergonomics for driving staff.

Finally, people love trams, sometimes for irrational but mostly for practical reasons.

4.8. Overhead line and electrification

The first tramways were horse drawn. A horse could pull 50 passengers in a tram, compared to only ten in a bus on a road. The impact of wars and equine illnesses made operators look for more reliable and economic alternatives to animal power.

With steam railways as the most obvious model in the nineteenth century, small steam locomotives were tried on various systems (Figure 4.55). The noise and dirt of steam locomotives in the public highway was not as acceptable as on remote rail rights of way. Other systems of motive power were also tried: batteries, cable, compressed air, clockwork and petrol engines. None were entirely satisfactory, although many cable-hauled systems were built around the world, including San Francisco in the USA, Melbourne and Sydney in Australia, Dunedin in New Zealand, and London, Edinburgh and Llandudno in the UK. Today, only San Francisco and Llandudno (Figure 4.56) still operate with cable haulage.

It did not take long for operators to realise that electric traction would offer many advantages. The first practical traction system was demonstrated by Siemens in Berlin, Germany, in 1882, which used the running rails to provide the electric power. In streets, this was not a practical option, so the overhead power line was first used by Siemens in Berlin. Siemens used a free-running trolley on the top of twin wires ('Troller'). This had serious practical difficulties, especially if more than one line was built, requiring the overhead line (OHL) to split or join at junctions.

Frank Sprague invented an upward sprung pole in the USA, in 1887, with a wheel pressed onto the underside of an insulated overhead cable. The practical demonstration of this was in Richmond, Virginia (Figure 4.57). The OHL trolley made current collection more reliable and junctions between lines became practical.

The conduit system was developed to avoid the use of OHLs. The electric conductors were placed underground, with vehicles collecting power through a slot in the road surface, normally between

Figure 4.55 Steam tram operation in Leeds 1885

Source: Alan Brotchie Collection

Figure 4.56 Cable tram in Llandudno, 2009

Source: Llandudno Council

Figure 4.57 Sprague's trolley system

Source: CS Dunbar

the running rails but sometimes under one of the running rails. This system of power supply found favour in several capital cities – for example, Brussels, Budapest, London, Paris, and Washington DC. Washington DC was the last operator to use the conduit system, closing in 1962 (Figure 4.9).

Apart from the high cost of excavating and installing the conduit, maintenance costs were also high, with debris falling into the conduit needing to be cleaned out, and at junctions if a vehicle came to a halt it lost power and was stranded. It had to wait for the next vehicle to push it onto the power rails on the other side of the junction. The conduit system, while practical, was very expensive to build and maintain. Most cities only had the conduit in the central area, with OHLs in the suburbs. This required change over points part way along routes, which held up services and were labour intensive. The delays could be severe, especially if there was a shortage of current collecting ploughs for streetcars heading to the downtown.

The second alternative, while using an underground electric cable, transmits power to passing streetcars by way of surface pads or plates. Various methods of activating these, including mechanical and magnetic, were tried but none proved entirely reliable, sometimes leaving the surface contact pads electrically live. Reports of electrocuted animals and the difficulty of maintaining these surface contact systems led to the first generation demise by 1910.

It is interesting, therefore, that Alstom and others have developed new surface contact systems. One was installed in the centre of Bordeaux, opening in 2005 (Figure 4.4). In this system, the surface plates are activated by a radio signal from approaching LRVs. At first this was unreliable but with more development has been made to work. The reported cost is at least four times as much to install as the OHL system. No figures on comparative maintenance costs are available. The high installation cost has led the system only to be used in central Bordeaux, with OHL elsewhere. There is, as yet, no evidence of a solution to the other problem which beset the earlier surface contact system, surface water from rain or snow leading to arcing or short circuiting the plates.

Other systems for powering tramways have included internal combustion engines, batteries and, latterly, flywheels to store energy on board. For various reasons, these have all proved impractical, unreliable or uneconomic. In future, with the development and reduction of cost of batteries or supercapacitors to store power and roof mounted photovoltaics, it is possible that autonomous power systems could be practical and not need an external power source. This, however, will remove one major advantage of an external power source which is the ability to draw on extra current, overload motors and get more power for acceleration and hill climbing.

Historically the overhead line became the universal method of electrifying streetcar lines and tramways but not before the other methods discussed earlier were tried. In the late nineteenth century, most towns and cities were already overhung with many wires carrying telegraph, telephone and electric power. Some city authorities were concerned that the additional tramway OHLs would be aesthetically unacceptable. There were also problems with interference of cables and reports that lighter telegraph cables were blown onto streetcar OHLs with dramatic and sometimes fatal results.

The OHL is the most practical and economic system for tramway and light rail electrification (Figure 4.58). OHLs can be aesthetically pleasing. With the advance of OHL design, these can now be inconspicuous, especially if adjacent buildings are used to support the OHL (Figure 4.59). Even if traction poles are used, these can be slender and in partnership with the highway authority used to provide for streetlights. The Vienna Opera House in a World Heritage City is surrounded by the OHL of the local tramway and indeed provides supports to avoid the need for poles (Figure 4.3).

There are three major considerations for the OHL system design engineer: electrical, mechanical and aesthetic. As LRVs have got heavier, more power (current) is needed for acceleration. External power supply also has one further advantage – it allows the vehicle designer to use small oversaturated motors during acceleration by drawing more power from the OHL, than the continuous rating of motors would indicate. Typically, this over saturation can be about 200% of the continuous rating. This is especially important for start/stop operation, when motors will be at full power

Figure 4.58 Typical power supply circuit

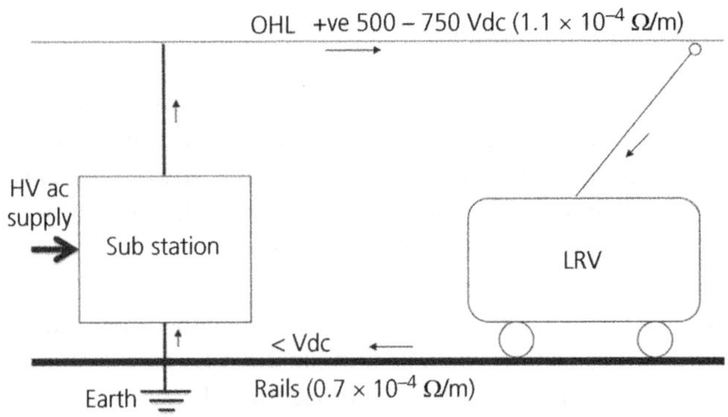

Source: The author

Figure 4.59 Span wire wall fixing, Sheffield 2008

Source: The author

for about 30 s. With careful attention to design of the traction control system, this means smaller and lighter motors can be used without overheating. Lighter motors reduce vehicle weight and, consequently, the need for power during acceleration. The motors can also be used in regenerative braking, which can be fed back into the OHL and power an adjacent tram. No system can be 100% efficient but reports of total power consumption being reduced by 30% are common.

Compared to the electrical resistivity of the running rails (10×10^{-8} Ω/m) used as the current return path, the OHL (1.7×10^{-8} Ωm) is about an order of magnitude less electrically resistant (Figure 4.58). In practice, because of the much larger cross section of the rails, the resistance of the tracks is about 40% lower than that of the OHL. In the UK, the voltage of the running rails should normally be less than 7 V above earth, requiring substations to be about 2 km (1.25 miles) apart. The voltage drop in the OHL over the same distance is likely to be about 50 V.

To reduce the power losses in the OHL, the electrical resistance needs to be reduced. This can be achieved by the choice of conductor, usually copper, copper alloy or aluminium. There are better metal conductors – for example, gold – but these are much more expensive and heavier. The second method for reducing OHL resistance is to use larger diameter cables, since the resistance is inversely proportional to the cross-section area but the diameter is proportional to the square root of the cross section, meaning that the resistance can be halved, with only a 40% increase in the diameter. This minimises visual impact compared to having a second wire to achieve the same reduction in resistance.

Another way to reduce the OHL resistivity, if there is a double track line, is to link the two OHLs to work as one. This does require connecting points, where the two can also be isolated, when for example work is being undertaken on one track and the OHL of that track must be unpowered for staff safety.

A fourth way to reduce resistivity is to have large section underground feeder cables in parallel with the OHL. This has the advantage that the combined resistance is less that the resistance of the feeder cable (Equation 4.3). A fifth way to reduce power losses in the OHL is by having more substations, or using low resistance cable to feed the OHL at shorter intervals – for example, 800 m.

Resistance of a combined parallel feeder cable for the OHL

$$1/R = 1/r_1 + 1/r_2 \tag{4.3}$$

where R = resistance of combined system, r_1 = OHL resistance and r_2 = feeder cable resistance

Let $r_1 = 50\,\Omega$ and $r_2 = 1\,\Omega$. So that $1/R = 0.02 + 1 = 1.02$ and $R = 0.98\,\Omega$

A sixth way may be possible for some suburban or segregated lines using railway style catenary, where the upper cable acts as the feeder to the lower contact wire. This is not acceptable in many city streets. Parallel power feeding to reduce losses on double tracks, with appropriate section isolation, makes it possible to have operational flexibility for maintenance and repair on one track, while still using the other, with temporary cross overs for contra-flow operation.

Whichever method of feeding power to the OHL is used, the major structural design task is the location of the support system, that will accommodate both the passage of light rail vehicles and cope with extremes of weather in terms of temperature (expansion/contraction), wind (sway and horizontal forces) and icing (extra weight and damage to pantographs or other current collector).

The OHL is normally suspended from span wires, or bracket arms from poles set into the ground. Span wires can also be directly hung from adjacent buildings (Figure 4.59), which reduces the need for poles and the visual clutter of streets – for example, the city centre section of the Manchester Metrolink was originally going to have 75% of the OHL directly hung from buildings and 25% from poles. In practice 75% is from poles, adding to street clutter, especially as 'I' section beams have been used for many of the supports. This has been a source of on-going criticism.

The pull-out force on a building support from double track span wires is quite low, usually under 1500 N (about 300 lbs), which masonry walls can easily accommodate. Historically a large plate (rosette) was fixed to the wall with four bolts cemented into the masonry. Today, a single 'eye' bolt is bonded into the masonry by an epoxy resin (Figure 4.59) or similar high-strength cement. The building owners are normally given a small annual payment and the LRV operator insures against building damage from the OHL.

Historically, OHL support poles have been sunk about 1800 mm (6 ft) deep in the ground, either between the tracks in the carriageway or outside the track area by planting at the kerb edge of the footway/sidewalk/pavement. Such poles usually have a slight rake (about 150 mm/6 in from vertical at the top) away from the track centre. This compensates for the weight of the wires. More recently, in the UK, heavy rail practice has crept in using concrete (gravity) blocks about 1 m (39.4 in) cubed set into the ground, with the poles bolted on top.

Finding the space for such large blocks is often difficult in the footway or sidewalk, with existing underground utilities, because the stresses in involved concrete blocks are over engineered. A demonstration project of a simplified OHL design was undertaken in 2004 at the Carnforth Railway Centre in the UK. Here, wooden poles 180 mm (7 in) diameter were set 1800 mm (6 ft) deep into non-descript mixed ground with very satisfactory results. Holes were dug with a small backhoe. The poles planted and the ground compacted, without the use of concrete. Within an hour of planting, linesmen were able to climb the poles (Figure 4.60) to fit bracket arms.

Light rail systems can use different OHL pole styles to suit local needs. The most widely used are made from steel tubes, often tapered or stepped. But there are also examples of concrete, timber, H-section rolled girders and lattice steel masts. There are also examples of suitable trees being used to anchor span wires. When using poles, they will be at least 8 m (26 ft) above the road surface, to allow for wire sag and to keep the OHL a safe height above the road surface. This is usually over 5.5 m (18 ft), with enough space for high (goods) vehicles and double-decker buses to pass

Figure 4.60 Wooden pole and bracket arm used at Carnforth

Source: The author

underneath and across the tracks. Some UK roads are reserved for strategic 'high vehicle' routes, where the OHL will need to be 6.5 m (23 ft) above the road. LRVs will therefore need 'high reach' pantographs.

Most light rail systems now use pantographs. The OHL, as well as following the line of the track, also wiggles from side to side across the track (Figure 4.61). This ensures an even wear on the pantograph head. For systems using a mixture of both trolley poles and pantographs, the pantograph will usually be mounted midway between bogie centres. Whereas on exclusively pantograph systems, mountings are above a bogie centre.

Normally, the OHL is double insulated from its supports. Given the low voltage involved, under 1000 V dc, simple insulators are normally adequate. Historically, these were ceramic. Today, one or other of the many plastics with the right mechanical and electrical properties can be used. For the Carnforth demonstration discussed above, plastic rope (Kevlar) was used as the primary insulator, with wooden poles, or plastic collars insulating the bracket arms from metal poles.

Mechanically fixing the OHL directly to span wires or bracket arms creates hard spots where pantographs (and trolleys) bounce, causing arcing, damaging the OHL and the current collector. Railways avoid this by using a second catenary support wire. This gives an elastic support to the entire OHL contact wire but then needs registration arms to keep the contact wire to the correct horizontal alignment. This is not practical or aesthetically acceptable for most street tramways. A new system demonstrated since 2004 at the Carnforth Railway Centre provides a fully elastic OHL suspension and zig-zag over track (Figure 4.61) without the use of a catenary or registration arms (Lesley, 2006).

Figure 4.61 OHL zig-zag for even pantograph wear

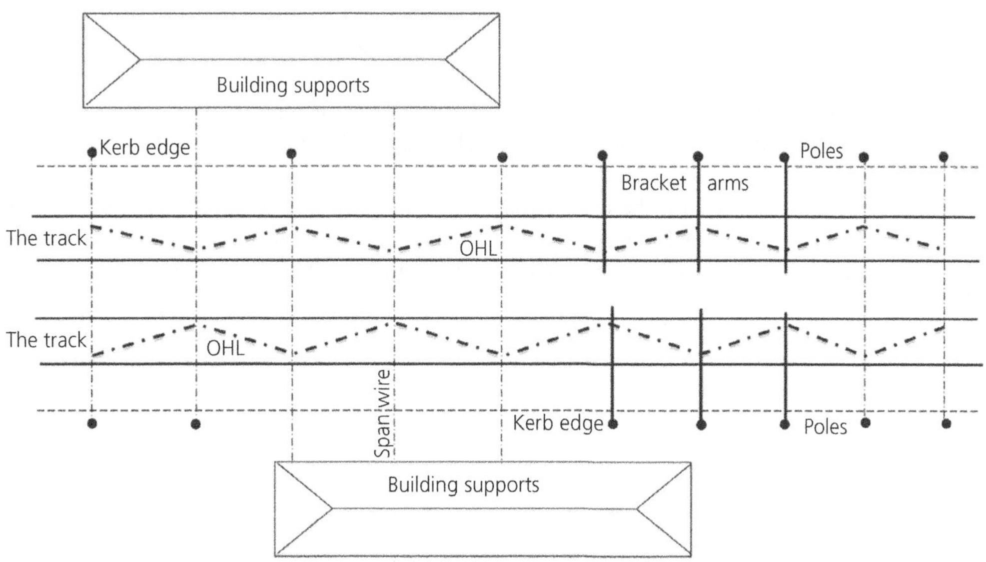

Source: The author

Historically, tramway OHL was not tensioned, being set up for average temperatures and, therefore, slack in hot weather (Figure 4.62) and taut in cold. Tensioning an OHL has many advantages, including temperature independent OHL height, consistent wear on pantographs and the ability to operate faster without OHL or pantograph damage.

Railway catenary tensioning uses weights or springs attached to the end of OHL cables. This needs OHL overlaps to maintain a conductor along the whole track. There are few examples of such tensioning in city centre streets. Most use un-tensioned OHL to avoid the clutter and pole strengthening needed to resist the typically 10 kN (1 t) or more force. The Carnforth demonstration patented system developed a completely new tensioning system which does not need OHL overlaps, nor large end tensioners, and places small extra forces on tensioning poles.

The relationship between the factors involved in hanging an OHL, following a catenary vertical curve between supports, can be found in Equation 4.4.

$$T = W \times L^2/(2 \times S) \qquad (4.4)$$

where (length) $L = 40$ m (130 ft) , (ag) $S = 0.2$ m (8 in) (weight) $W = 54$ kg (120 lb) then

(tension) $T = 54 \times 40 \times 40 / (2 \times 0.2) = 216$ kN (500 lb force)

If $S = 0.01$ m (4 in) then

$T = 54 \times 40 \times 40 / (2 \times 0.1) = 435$ kN (950 lb force)

This allows the overhead design engineer to look at the option of longer span lengths on straight tracks, against higher tensions to maintain an acceptable sag.

Thus, $L = 50$ m (164ft) $T = 844$ kN (1860 lb force)

Curves present another challenge for the OHL design engineer, to maintain alignment over the track, accommodating the dynamic kinetic envelope of passing light rail vehicles and keeping pantographs in contact with the OHL, whatever the weather conditions. Small curve radii need care with the design, since fluctuating temperatures can change the alignment of the OHL.

This can be accommodated with the use of span wires and pull offs. Figure 4.63 shows a 25 m (80 ft) radius curve on street track. The OHL engineer has used split wire supports to prevent sharp kinks forming in the OHL, which could bounce the pantograph and cause sparking, damaging the OHL and pantographs.

Figure 4.62 OHL vertical sag

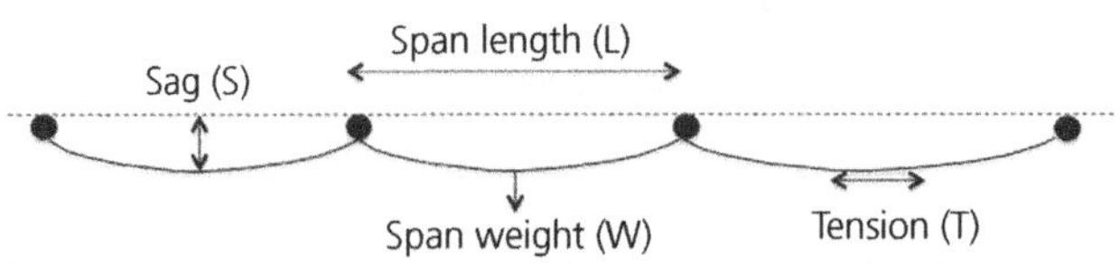

Figure 4.63 Aligning OHL over curved track, Sheffield, 2008

Source: The author

In between the pull offs, the OHL follows a straight line, so the points of OHL support on curves are usually on the outer edge of the pantograph track. The distance between two supports allows the OHL to migrate to the extreme points of contact on the pantograph. This also takes into account the pantograph and vehicle sway. It is worth noting the minimum number of traction poles required by this arrangement and that all the poles are on the outside of the curve, with adjacent poles sharing the middle OHL support.

Finally, for the new system demonstrated at the Carnforth Railway Centre, differences in ambient temperature and, hence, OHL length are compensated by the pull-offs raising or lowering the contract height of the OHL. This is thus a self-correcting temperature compensating system that always has the right horizontal alignment for complete pantograph running.

4.9. Sub stations and distribution

The first generation of electric street railways and tramways, in the nineteenth century, generated their own power because national electric power supply was still in its infancy. Most cities did not have electrical grids (Figure 4.64). Later, as tramways were municipalised, the power station formed the basis of an urban distribution network for domestic electrification.

Today, most light rail systems buy their power from central or public generators. This will be connected to the light rail system at a high voltage, normally 10–25 kV three-phase alternating current. Substations about 2 km (1.25 miles) apart then transform and rectify the power to the OHL line voltage, usually 500–750 V dc.

Substations are normally unmanned and are remotely controlled including the ability to switch power on and off. A substation can be isolated with a section of track served, especially in an emergency. Inside the substation there will be considerable protection, including high-speed circuit breakers, in the case of a short circuit on the line, or inside a light rail vehicle. Usually these operate

Figure 4.64 Nineteenth century substation in Leeds

in milli-seconds. Unlike ac power, dc will create arcs when switched off. Mechanical circuit breakers must therefore be able to open very rapidly. Solid state switch gear can open circuits even more quickly and without the danger of creating arcs.

Remotely monitoring the functioning of substations is now standard, either using a dedicated cable network (wire or fibre optics), radio links through the air, or a radio frequency carrier over the OHL.

Stray currents (Figure 4.65) have been raised as an issue in new UK systems, causing possible damage to under street utilities and steel foundation frameworks of buildings. By and large in most countries, stray currents are only considered a problem when actual damage (e.g. steel corrosion) is found. In the UK, measures to prevent strays have added considerably to the cost and complexity of the tracks of new light rail systems.

These measures have included the use of Faraday cages under the track, which one senior engineer has claimed are ineffective. Track was relayed in Easter 2007 by the same engineer on the Croydon Tramlink, omitting stray current protection, based on cost savings, and found that there was no observable stray current effects. Two other methods of stray current reduction are available. One systemic and the other location specific.

If the track, for the return circuit, is not earthed, then there is little scope for currents to stray through the earth. There are unearthed light rail systems that are as safe as the majority which are

Figure 4.65 Stray current paths

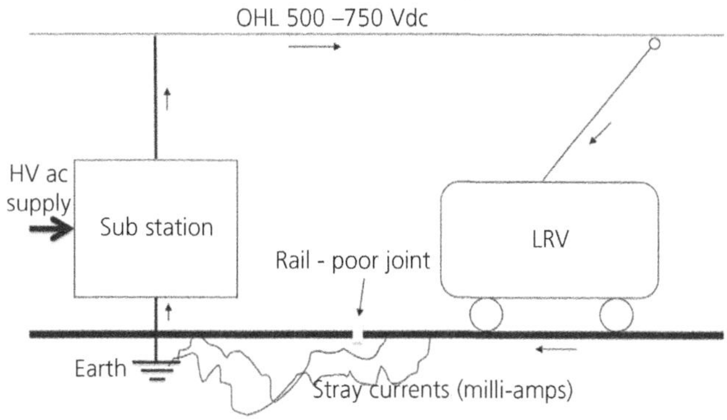

Source: The author

earthed. Another way to reduce stray currents is by totally insulating the rails from the ground. Embedment polymers are a widespread method of insulating tracks, mechanically as well as electrically. In wet weather, all earthed systems will have stray currents. These will result from water lying on the road and over rails, with currents going across the road surface and down drainage channels and so on, following the rainwater. This will be especially bad if there are breaks in the track, preventing currents from reaching substations directly.

Location-specific measures include diodes to prevent catalytic corrosion of steel work. This, however, requires the excavation and treatment of the steelwork. For most of the world, outside the UK, this is the protection most widely adopted.

Unmanned light rail substations typically with a 500–1500 kW capacity can be fitted in small spaces, including basements or other places which would normally not be in active use. Prefabricated portable containers can also be used. Substations will have a connection to an incoming high-voltage supply. This will include breakers, fuses and isolators and may include metering for payment.

Transformers and rectifiers are now compact, leaving only the dc switch gear to isolate the OHL from the substation, or to isolate just one OHL line for track or other maintenance, while allowing the other track to be used safely by service trams. Switch gear for dc used to be mechanical but is now usually solid state and, therefore, compact. Safety monitoring will also be linked to the central control, so the control engineer can see the health of all the equipment and take preventative maintenance steps, where needed. Substations will also have monitoring micro-processors, that will flag up any faults and help prepare maintenance schedules.

Cables from the substation supply feeder boxes about every 800 m (900 yds). These also perform OHL isolation functions. These parallel feeder cables will maintain OHL voltage when LRVs are drawing power, by reducing the overall resistance of the OHL network as discussed above (Equation 4.3). The ducting for these cables can be a considerable expense, unless undertaken at the same time that tracks are laid. Cables in these ducts will be specified <3000 A and will be installed so that there is sufficient cooling to prevent the cables overheating. This is important, as overheating increases the resistance of the cables and, therefore, the power loss in the system. Overheating is not normally a problem for light rail systems, as power drawn by passing LRVs is variable, transitory and rarely at maximum current for more than about 30 s, except perhaps for key feeders in busy city centres or up long or steep hills. The power needed for cruise speed is orders of magnitude of that needed for acceleration (Section 4.4.2).

Some LRVs regenerate current on braking and feed that back into the OHL, if it is receptive – for example, another LRV in the same section drawing current during acceleration. The regenerated current can be conveniently used in batteries, either on board vehicles or in substations. There will be a maximum voltage specified for the OHL. Should regenerated power exceed that, then the power may be consumed in the LRV – for example, in heating resistors.

On a light rail system with an operating service of about 5 min or longer, the probability of this is low. In this case the voltage of the OHL may rise dramatically to over 1000 V and damage safely equipment and switchgear. Substations are therefore fitted with a way to absorb regenerated current – for example, inversion to ac and then feeding back into the supply grid. With solid state equipment, this is practical and efficient. Given the small power involved however the capital cost may be uneconomic.

There have been proposals for fitting energy storage flywheels in substations, which as well as absorbing regeneration currents can also provide voltage stabilisation. Similarly, banks of batteries can perform the same function more efficiently but at a higher capital cost. No such installations are, as yet, widely used. The development of super capacitors and their reduction in cost can achieve the same result. Nevertheless, the maximum power saving is well under 50%. This saving needs to be off set against the increased capital and maintenance costs of more complicated substations, not to mention the LRVs.

Figure 4.66 Contemporary substations

4.10. Conclusions

This chapter has considered all the engineering issues that need to be carefully designed at an early stage of the project. This is when conflicts can be identified and rectified before any physical works have begun. Getting the engineering right first time is the most cost-effective approach. Some innovations can be separately tested to ensure that the theory translates into practice. These steps will also create a high level of confidence, both with the decision makers and the public, who must be persuaded to ride the trams, justifying the system investment. This is particularly important when car trips are to be attracted from and the level of traffic congestion and pollution decreased.

LJS Lesley
ISBN 978-1-83708-399-2
https://doi.org/10.1108/978-1-83708-398-520261006

Chapter 5
Engineering practice

This chapter discusses the challenges of translating engineering design from Chapter 4 into practice on the ground.

5.1. Introduction

Taking the conceptual drawings and plans into reality and everyday operation is a large step. This chapter sets out most of the key stages needed. These include the maintenance of the system once it is running to ensure safety and reliable service. Over the life of a tramway/light rail system, maintenance may cost many more times than the first (capital) cost. Constructing and equipping a reliable system should be a high priority for the design engineers, to minimise the need for later maintenance, or worse retro replacement.

5.2. Construction and installation

The most visible part of installing a new light rail system is construction in streets. Great care in planning is therefore needed together with good engineering design, in order to minimise public impacts, including traffic disruption. One of the major criticisms of the construction of the Sheffield, England, system in 1993–94 was that the contractor treated the whole city as a construction site. This caused considerable disturbance to traffic flows and to bus services. Businesses and frontagers faced a decline in trade, as customers had difficulty reaching premises. Similar problems were experienced in Edinburgh, where track construction began in 2008. The first phase opened in 2018 and the second in 2023.

The largest part of the preparation for the construction of a new light rail system, which is not on segregated rights of way, will be the roads where tracks are embedded. This may include the relocation of under street utility plant, especially where the track is physically in the way of access. In many places, plant may become inaccessible, so it has to be relocated. Leaving plant in place is the preferred option of many utility companies, provided safe access arrangements can be agreed with the light rail operator, for plant maintenance and repair.

The use of concrete track foundation slabs, can either be laid by slip form paviour, or cast in-situ in sections with shuttering. This further extends the period of disruption while the concrete cures to the required strength. Only one system using precast concrete track panels has been widely used, developed in Budapest (Figure 5.1) and widely applied in the Soviet Comecon countries until 1990 (Horvath *et al.*, 1977). The panels suffered from poor cohesion with the substrate, leading to heavy vehicles rocking and other failures.

A variety of alternate street track forms have been proposed using grooved girder rails. As yet, none have been applied to a live street tramway (Lesley, 2023). Since 1996 a new track form has been in service, without needing any maintenance, on the South Yorkshire Supertramway. This uses a new low-profile rail, provided by British Steel and narrow (360 mm/14 in) precast shallow

Figure 5.1 Schematic of Budapest prefabricated slab track

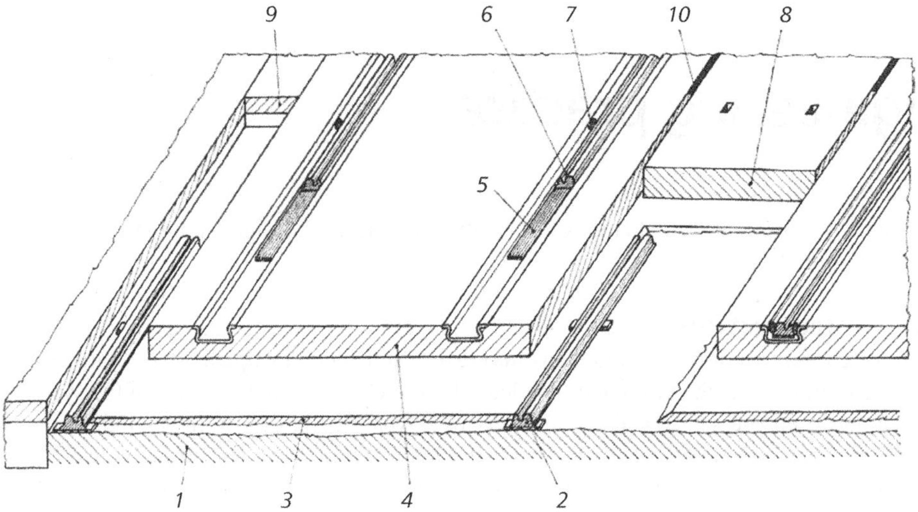

Source: Miklos Bagi

Figure 5.2 LR55 Track in Sheffield and tested in Rotherham Bus Station

Source: The author

(180 mm/7 in) rail foundation beams (Figure 5.2). This has been fully described in a number of publications (Lesley et al., 1998). It replaced girder rail track, failed in less than 6 months of service from the impact of heavy goods vehicles.

Constructing tracks on segregated rights of way is easier for the engineer, since there will be exclusive use of the space and materials can safely be stored on the site without interfering with the public. There are, however, challenges in constructing on continuous and narrow strips of land, including lack of side access. A 'moving front' strategy can be adopted, with deliveries using the completed rail sections.

5.2.1 Track works

The design engineer's challenge must be a least-disruptive construction sequence, including the delivery of materials to site. Unlike many construction projects, installing rail tracks in city streets is a very public activity and rarely can 100% occupation be possible, even if all vehicular traffic is removed, since the city has to continue to function and frontagers along the route need access to their property, as do pedestrians.

Lessons should be learned from the tramway installation in Sheffield 1992–1994, where the contractor effectively treated long lengths of road as a conventional construction site, with quasi-mainline railway occupation. These lessons unfortunately were not applied to the construction of some of the Edinburgh tramway 2008–2023, which caused considerable public protest and the opening of the first phase delayed until 2018. The second phase was opened in 2023.

The first lesson is the need for proper consultation with frontagers and the public in general, to find out how best to maintain access and minimise environmental and other disruption. In residential areas during the critical weeks of track installation, the contractor could offer the residents free holidays, with suitable property security when vacant. This avoids inconvenience for residents and can be a 'thank you' for their co-operation. Not having to work around people can reduce installation costs, to offset the cost of offering residents holidays. This should also make for a faster and neater installation. Contractors preparing to install LR55 tracks claim to be able to lay complete 100 m (328 ft) per week, without complete street closure. In comparison the London County Council in 1904 using manual labour and no machines was able to install 91 m (300 ft) per day.

The most disruptive operation is installing the tracks and, where needed, relocating under street utility plant. Good consultation with the utility companies will pay handsomely, since relatively minor redesigns of track alignments can produce very substantial cost savings by reducing or avoiding the need to divert plant. Even larger savings are possible by a minimum disruption of the highway. Figure 5.3 shows an installation sequence from 2009 track relaying in Budapest.

Figure 5.3 Budapest embedded track laying sequence 2009. Track on strip foundations. Mass concrete poured sand for road blocks

Source: The author

Inevitably, the tram track horizontal and vertical alignments may not be identical with the existing road. Traditional urban streets are normally profiled with a crown along the centreline, so that rain water drains sideways into kerbside gullies and drains. Tram tracks are normally laid horizontally or nearly so, so that passengers have a comfortable ride and the wear on wheels, bearings and rails is equalised between the two sides of vehicles and track.

Should the highway pavement be reprofiled to the ultimate alignment of the track, before or after installation? Most track installers have gone for post-track installation road surface reprofiling (Figure 5.3). The benefit of a pre-installation reprofiling is that it is a simple highway procedure, with many competent contractors able to give competitive quotations.

Not having to work around rail tracks also makes the resurfacing less complicated. Pavement reprofiling after the tracks are in place is more complex, with an interface between road and rails. There are fewer contractors competent or willing to undertake such works, at a competitive price, since there is much scope for contracts to go wrong.

By reprofiling the road surface first to the ultimate track alignment, the design engineer can undertake quality control in two stages. The first being the highway pavement, where if any remedial measures are needed, they are clearly down to the road contractor. The second is with the track installer, who can use the highway pavement as a datum. Thus, any errors in the track profile are down to the track installer. The installation process should also be simplified, as the two operations are separate and in sequence. Installing tracks when under-street utilities need to be relocated means extensive excavation and so post pavement profiling is the only option.

Installing tramway or streetcar tracks will depend on the kind of rails used. Grooved girder rails – for example, Ri60/60Ri (Figure 5.4) which are 180 mm (7 in) high and bottom supported – require a foundation beneath the rails. Recent new systems in France have effectively used mainline railway track designs with a crushed rock ballast base and the rails clipped in place on sleeper/ties. This means that a width of about 6000 mm (20 ft) of the road has to be excavated to a depth of at least 1000 mm (40 in).

Figure 5.4 Ri60/60Ri and LR55 rail profiles compared

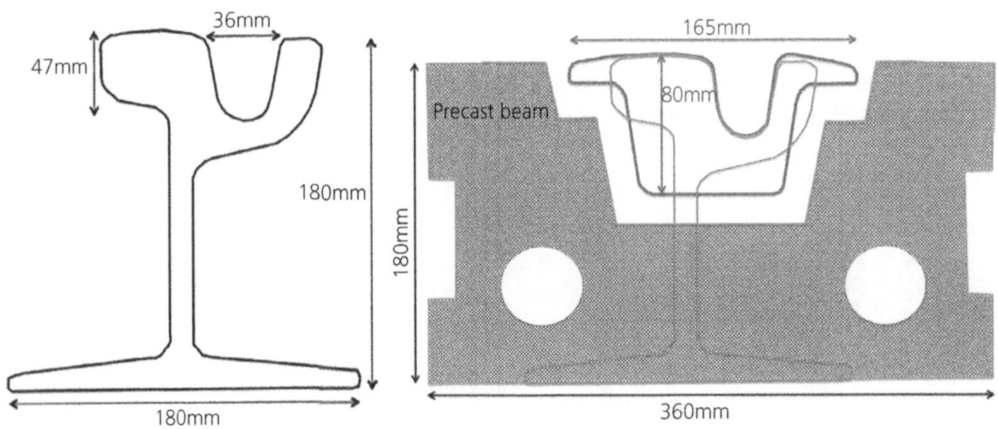

Source: The author

The original rail profile may not be available for extensions to existing systems. This then needs a transition between different profiles – for example, using low profile LR55 rails (Figure 5.5) to extend existing 'Phoenix' grooved tracks, like Ri60/60Ri, will need a very different transition (Figure 5.6), not only to match the different profiles but also to maintain rail stiffness across the transition.

Many North American and other European systems use a concrete slab between 300 and 500 mm (12 in–18 in) thick as the track foundation, with the rails clipped or bonded on top. This needs an excavation depth of between 500–700mm (18–28 in), excluding any utility relocation. The LR55 system developed by the author uses the strength and stability of the existing road pavement as its base, with shallow prefabricated concrete foundation beams only 180 mm (7 in) deep and 360 mm (14 in) wide, set into narrow recesses 200 mm (8 in) deep and 400 mm (16 in) wide. This minimal excavation is possible by the use of a top supported rail only 80 mm (3 in) high (Figure 5.4).

Figure 5.5 LR55/Ri60 transition rail detail

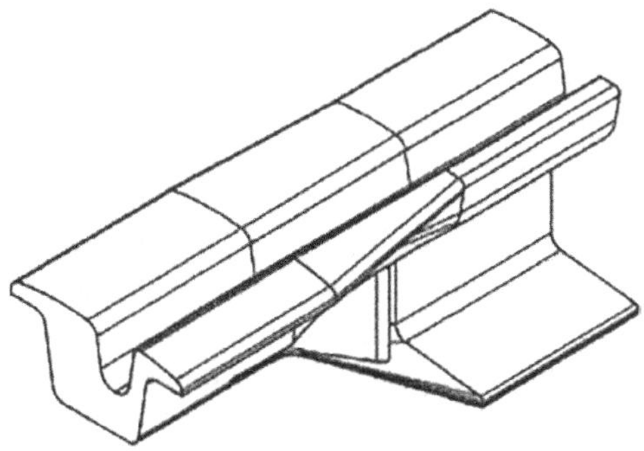

Source: David Turner, Direct Track Solutions

Figure 5.6 Embedded 'scissors' crossover in Budapest 2009

Source: The author

Probably the most complex part of this process is the installation of pointwork/turnouts; switches and crossings, where tracks divide, merge or cross over each other. The design engineer is responsible for ensuring that the track fabricator has the horizontal and vertical track plans. This will enable the fabricator to pre-assemble the track components off site, to ensure that the fit to plan is accurate and to make any adjustments prior to reassembly on site.

Light rail vehicles traverse street-based switches and crossings at low line speeds due to the small radii employed. This means that passengers are subjected to further braking and acceleration, and increased power consumption. The reason for this is that street track switches usually have a radius of about 20 m (66 ft), without transitions, so that for the comfort of passengers, limiting horizontal and centrifugal jerk rates means slowing down, often to as slow as 5 km/h (3 mph). Where space is limited – for example, at terminals – scissors crossings (Figure 5.6) can be used, since this provides flexibility in the use of the terminal tracks and stop platforms.

New ideas for switches can avoid both the potential discomfort of short radius curves and the need for trams to slow down unnecessarily. This can be achieved using 'pre-sorting' points. Here the switches are set back some distance from the junction (50 m or 160 ft) using large radius switches >200 m (650 ft). Two tracks are then interlaced, running side by side until the junction. At the junction the diverging track can include a proper horizontal transition, with only a simple crossing, where derailments are practically zero.

There is a second benefit of pre-sorting junctions, where LRVs have traffic signal priority. This makes co-ordination with the traffic signals easier and thus providing a 'green wave' for trams, including those turning from the main line. This is the norm in many European cities. Where there is not enough room for a pre-sorting arrangement, use of transition curves across the switch blades starting with radii greater than 100 m (325 ft) and reducing to 25 m (80 ft) will be more comfortable for passengers and less damaging on track and vehicles. Here, close co-operation between the design engineer and trackwork fabricators will pay dividends. This should mean less maintenance. Where there is not room for this, then an initial 25 m radius followed by a straight connection to the second track, with a reverse 25 m radius, will achieve a similar comfortable result.

Historically, embedded road switches and crossings created additional and often very annoying wheel/rail noise, particularly as light rail vehicles are flange guided around short radius curves. The 'clunk' of wheels over the crossing gap is a noise nuisance. The use of pre-sorting points means that one source of noise is reduced, as well as reducing switch blade wear and the need for maintenance. The second noise source, common in railways, is the crossing gap. This can be avoided by the use of 'silent' crossings, which evolved well over 100 years ago (Figure 5.7). Here the bottom of the groove is raised at the intersection of the rails, so that wheels are momentarily flange running, rather than tyre supported.

A variation of this is where the groove on the parallel rail is also raised, so that both wheels are flange supported, thus reducing any extra torsional forces in axles when wheels are solidly attached to the same axle. Such stresses over 'silent' crossings are, however, small compared to traversing short radius curves. Historically, there is little evidence of any problems arising from this.

Most European tramways have used 'silent points or turnouts' for over 100 years. These give satisfactory service and are environmentally benign. Being good neighbours from less noise without loud 'clunks' is a nuisance that can be avoided. The first generation of UK tramways all had silent

Figure 5.7 'Silent' crossing with worn raised groove

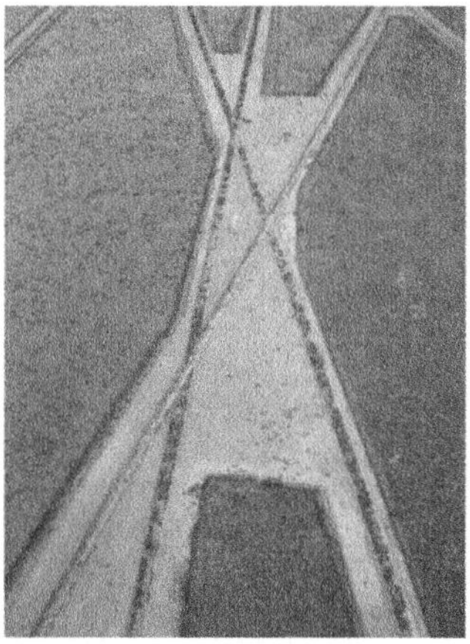

Source: The author

points. The second generation have been influenced by railway practice using noisy open crossings. This is a specific example where the design engineer faced with a technical problem would do well first to search the literature (or websites) to find if it has previously been solved. This can save significant time, cost and heart ache for the track designers, system users and those living or working in the area.

Return current cables are connected to feeders and substations. This will ensure that any rail joints not welded will have electrical continuity. Any vehicle detection equipment can also be installed at junctions, as well as cables for the operation of switch motors.

Making up the highway pavement is the final part of embedded track installations. Tarmac or similar flexible pavement materials are the easiest to use. These can be rolled between and outside the rails, to be level with the top of the rails. This gives a smooth road surface and a good waterproof seal against the rails. Some cities use stone blocks to make up the road surface, which are sealed with tar, bitumen or other mastic. This has the advantage of being easily removed for track maintenance but provides an uncomfortable ride for other road vehicles. This might make sense where tracks are exclusive for light rail vehicles or in historic town centres, to discourage use by motor vehicles.

Segregated tracks that are not embedded in highway pavements have a choice of track form. Construction is more straightforward. This is because conventional flat bottom rails simply supported on sleepers/ties set on crushed stone ballast is the most widely available and, therefore, lowest cost option (Figure 5.8).

Having cleared the right of way and protecting any crossroads to properties and for other needs, the first stage will be the removal of top soil. If the tracks are subsequently grassed, some of the top soil can be kept for this. The rest must be disposed of. In urban areas, normally there is a market for the supply of top soil for gardening and horticulture and this might generate a little income to offset part of the cost of installing the tracks, providing the administration is not costly.

Figure 5.8 Example of segregated open light rail tracks (Blackpool)

Source: The author

Below the ballast is the subsoil. Drainage must be provided, since water logging is a common cause for tracks to fail. This can be by grading the formation into side drainage ditches, or incorporating perforated drainage ducts along the line of the tracks. A geotextile barrier will be placed over this to keep the subsoil and ballast separated. Then a ballast of crushed aggregate about 25 mm (1 in) in size acts as an elastic drainage base for the tracks.

Pre-fabricated panels of track 18 m (60 ft) long can be brought to site, laid and aligned. The rails can then be welded in situ into long strings. Otherwise, pre-welded long rail strings are brought to site to replace the existing rails. Once the rails are at the correct line and level, further ballast is laid between the sleepers to anchor the tracks laterally (Figure 4.17). If a softer appearance is desired, a further geotextile sheet is laid over the ballast and top soil spread and seeded (Figure 4.21). The geotextile prevents the top soil contaminating the ballast, since its elastic properties depend on the ballast remaining clean and free draining.

5.2.2 Overhead lines

Some aspects of this have been discussed in Section 4.6. Electric tram and light rail operation during the last 140 years have seen numerous attempts to supply the power to the vehicles without the need for overhead lines (OHL). Of these, the conduit system with a continuous slot between the running rails and electric supply rails below the street in a hollow duct was the most widespread and reliable. To save having wires above city streets, conduit tracks of great complexity, cost and operational difficulty were installed in many cities, including Brussels, Budapest, London, Paris, Vienna and Washington DC (Figure 4.9). Washington DC was the final operational system, closing in 1962, ten years after the last conduit powered tram ran in Europe (London). While this avoided the 'wires in the sky' it meant more metal in the road surface. The friction of 'plough' on the slot sides increased power consumption and operating costs. This first cost was some ten times higher than OHL with much higher maintenance costs. Such systems were therefore constrained to urban centres, with OHL used in the suburbs, requiring change over points.

Other historical systems, including having electrical conductors on the road surface, all failed for technical, safety or cost reasons. Recently, a new surface contact system has been demonstrated in Bordeaux by Alstom, and Bombardier has developed another system relying on inductive transfer of power between under street cables and inductors on vehicles. These other systems are more complex and expensive to install and maintain than OHL.

OHL remains the most reliable, fully developed and available power supply system, which is not proprietary based. The application of good design means that OHL can be aesthetically acceptable even in World Heritage cities like Vienna. OHL is quick and economic to install and maintain. Installing OHL in city streets requires planning like the track works. The function of the OHL is to supply electric power to LRVs safely, while at the same time being aesthetically pleasing and environmentally unobtrusive.

The pre-planning survey will include locating and identifying street furniture that should be avoided or may need to be moved, and suitable buildings that can be used for the direct suspension of the OHL (Figure 4.5) and thus avoiding the need for traction poles. Faster and heavier LRVs draw large currents, so need bigger or more supply cables to minimise power losses.

Designers should use an elastically suspended OHL. This avoids hard spots where pantographs or trolleys can bounce, causing arcing that will damage both OHL and pantograph. The ultimate elastic suspension is a full catenary. For off-highway locations this might be the most appropriate solution, especially as the second (catenary) cable can act as a feeder for the higher currents drawn, when vehicles accelerate.

In city streets, however, catenary is normally unacceptable due to visual impact and increasing the clutter of the streetscape, although cities like Manchester and Sheffield use twin cables suspended side by side as a compromise, for supplying the high currents needed, with an acceptable OHL resistivity.

The engineering design for the installation of street traction poles has nearly been lost in the US and UK, during the 50 years since streetcars and tramways were abandoned. The horizontal forces exerted by span wires supporting the OHL are low, typically about 1500 N (about 300 lb force). This can be compensated for by erecting poles with a slight outward lean (about 150 mm/6 in at the top). The OHL weight supported by the span wire will pull poles inward to vertical.

The same logic applies when bracket arms (Figure 5.9) are used instead of span wires. The downward weight imposed on the bracket arm by the OHL will produce a turning moment balanced by the bending of the pole back to vertical, unless centre poles are used (Figure 5.8). In this case the weight of two OHL cables balance each other and the net turning moment on the poles is zero.

Planting OHL traction poles was, until recently, achieved by the use of an auger, to make a hole about 1800 m (6 ft) deep, slightly larger than the pole base diameter, and then using a cement/concrete slurry to seal the pole in place in the ground. After a short period, poles will be firm enough to string the OHL. The most recent UK new tramways have, in many cases, used traction poles significantly over specified and bolted to the top of foundation 'gravity' concrete blocks, about 1 m cubed, more appropriate for main line catenary. These will further complicate the installation, since the pavement/sidewalk must be cleared of utility plant.

Figure 5.9 Cable stayed side bracket arms, Carnforth Railway Centre

Source: The author

The retired engineering director Bill Gibson, of Blackpool Tramways (with unbroken electric operation since 1885), relates how the OHL designer for one new UK system visited Blackpool to inspect the OHL. Looking up he is reported to have said 'according to my calculations that is impossible'. The Blackpool engineer replied that this design has not failed in over 100 years.

Experiments undertaken at the Carnforth Railway Centre determined how strong simply planted traction poles can be. The poles were erected by United Utilities, the local electrical power supplier. Plain wooden poles 9 m (30 ft) long and 180 mm (7 in) in diameter were used. An auger proved to be impractical because of the very variable nature of the ground, which included areas of quarry rubble and waste steam locomotive fire ash. The holes were excavated by a small backhoe (mini JCB) and, once the pole was in place, back filled and compacted by the backhoe. Within an hour, the poles were stable enough for an installer to climb up the poles with spiked boots and belts, to attach the bracket arms (Figure 4.60). Where the ground was especially weak, a cement-based ground reinforcer was added to the backfilling, which overnight created a stable pole.

Historically, wooden poles and indeed trees have been used to support the OHL. Tubular steel poles are often used, sometimes filled with concrete after planting. The kind of poles used will depend on economic, engineering and political considerations. For street traction poles, careful planning is needed to locate the optimal places for poles, to support the OHL to the correct alignment for track geometry, vehicle swept path and pantograph position.

Existing street furniture, including lighting columns, will also have to be considered. Substantial lighting columns could be used for the OHL, except they may lack strength or bend inwards, so might need replanting. Alternately, the new traction poles could be used to suspend street lighting fittings to reduce the clutter of poles (Figure 5.10). In any case, consultation with the highway

Figure 5.10 Shared traction and lighting poles, Liverpool 1938

Source: Liverpool City Engineer

authorities will be needed to ensure that road signs are not obscured. In conservation areas and areas of special historic or architectural interest, the planning authority will also require very close consultation, so that poles and location are sympathetic to the local area.

During the actual installation in street, temporary local fencing will be needed to enable the pole hole to be dug, while protecting the public in the vicinity. Depending on location, a series of holes can be prepared, so that a truck with the poles, using a Hiab or similar hoist, can proceed along the street and plant many poles in one pass, minimising local disruption as well as speeding installation and minimising costs.

In many city streets, stringing OHL from adjacent buildings is a more environmentally pleasing solution (Figure 4.59). Older masonry, brick or stonework buildings are usually strong enough to support span wires. Historically, elaborate plates (rosettes) were bolted to building walls, to which the span wires were then fixed. Synthetic resins are today mostly used to fix anchor bolts into walls. This means that structural bolts are unobtrusive and economical, although without any architectural character. Many modern buildings have steel or concrete frames, with light weight claddings. The claddings are usually unsuitable for attaching span wires. Where the building's frame can be accessed, this will provide a suitable support for span wires.

In order to fix OHL span wires to buildings, the agreement of the owners is required. This is another area for early discussions, to generate good public relations. Building owners agreeing to OHL fixings is much better than using compulsory powers. Compulsory powers will generate negative publicity for the promoter of a new system or even outright popular resistance. This happened to the proposed West London tramway. The political party which promoted the scheme lost their representation in two out of three districts concerned, and the scheme was abandoned.

Installing building fixings, as well as bracket arms on traction poles, will normally be achieved using a 'cherry picker' or similar mobile high reach platform (Figure 5.11). One of these was used in the Carnforth OHL project described above, to install the bracket arm assemblies, as well as the cable. Using self-powered work platforms enables installation crews to proceed quickly and safely along a street, installing the OHL fixings on a moving front basis.

The actual location of the OHL will depend on the track geometry and LRV characteristics. For new systems, the OHL will follow the centre line of the track, with a horizontal 'zig zag' to ensure even pantograph wear. A short-wave zig zag between adjacent span wire attachments (Figure 4.61) needs no bracket arms to keep stable. On double tracks, the OHL should zig zag in opposite directions, so that the lateral forces on the poles or span wires balance each other.

On curves, the OHL will need pull outs to maintain the centre line within geometric tolerance of tracks, vehicle swept paths and dynamic kinetic envelopes. These pull-outs can be supported on span wires parallel to the track, to minimise the number of traction poles or from supplementary building supports used (Figure 4.63).

Tensioning tramway OHL can be a 'black art'. Some systems do not have active tensioning and operate by self-tension, created by the weight of the OHL with variable sag, depending on the ambient temperature. (Equation 4.8). For systems using active tensioning, this is relatively simple for off street tracks. Tensioning weights, springs or hydraulic rams are well established and readily available. The calculations needed are easily applied. Tensioning on street is more difficult, since

Figure 5.11 'Cherry picker' used for OHL installation

Source: The author

existing tensioning systems are bulky and not conducive to a simple aesthetically pleasing appearance. This is one reason why tensioning has been rare for streetcar systems.

A new simple and unobtrusive tensioning system has been developed which can be incorporated inside hollow traction poles and is easy to install. This patented system is available under licence to all competent OHL installers, and has been tested since 2004 on the Carnforth test installation.

Installing the OHL is the trickiest part of the process, since, ideally, it should be put up right first time with the minimum subsequent adjustments. The OHL will be installed on a moving front process. This starts with anchoring a working end, either to pre-installed OHL or a traction pole or similar fixing point. In attaching subsequent fixings to span wires or bracket arms, the correct horizontal location above the track and height above the road, the working tension can be achieved simultaneously. This is the most skilled part of the process. If a new team is being used, it would be best to practice it first at an off-street location.

Normally, a template sat on a track trolley will show the correct line and compensate for any cant or change in horizontal track alignment. It will also mimic the LRV and its pantograph trajectory. Other aspects of OHL erection will need a supervising engineer to be present. Space available in a cherry picker platform may not allow this for the final clipping of the OHL. As the OHL will be supplied in about 1000 m (3400 ft) lengths on a drum (Figure 5.12), maintaining working tension and alignment will be critical to the final alignment and performance of the OHL.

For the Carnforth project, the OHL cable was first run out over snatch blocks on alternate poles over the length of the track. This was less problematic to the final clipping than having to run out the OHL and, at the same time, clip into position to the right tension. In a street tramway with curves and corners, further complications will have to be accommodated to ensure that the correct alignment is maintained. This might also be by way of temporarily snatch blocks, looped over span wires or bracket arms. Either way, this is a two-pass process, in which in the second pass is the final alignment, when height and tension are achieved. As the clipping progressed, a mechanical tensioner was used at Carnforth to provide a tension of 30 kN (3 tons) in the OHL.

Figure 5.12 The author helping to run out OHL at Carnforth

Source: The author

The final check is a light rail vehicle moving at low speed with its pantograph raised, to confirm continuous contact with the OHL and that the contact pressure is uniform along the whole line. Minor adjustments to correct small out of design variations can easily be undertaken from a mobile work platform, progressing with the LRV.

5.2.3 Stops

In engineering terms, the track and OHL are major design and installation challenges. For passengers, however, the most important aspect of the system is the stops where journeys start and end. Next in importance are the vehicles on which journeys are made. These are the main and direct passenger experiences of using a light rail system. It is, thus, important to have good stop designs, constructed to the highest standards, within the budget for a financially viable system.

It is also important to consider local residents, who may not become passengers. Stops are a permanent part of the townscape and will be a neighbour to dwellings or businesses. Stops also affect the efficiency of the light rail system and add to the civic status of the area. Stops built to high design standards can be good ambassadors for the light rail system, creating a strong corporate identity. A good stop design will have a friendly ambiance. This is important both to attract passengers who are private car commuters and also to become a good neighbour for local residents. The use of a local architect can also enhance the acceptability of tram stops.

The first journey by light rail may be unfamiliar and create passenger anxiety. Car drivers used to getting in their car and driving will be especially nervous in having to wait at a stop with strangers. The perception of private car travel is that journeys can be completed non-stop. The great attraction is that a car is always ready. Traffic jams are lived with, or over looked, compared to the benefits and convenience of controlling personal travel.

Tram stops, where passengers interchange lines, can be particularly unnerving to new passengers, even with excellent signing, since people do not want to appear novices, or worse get on a vehicle going in the wrong direction. Most of the above is particularly relevant to influencing the design stage. Construction and fitting out must also be carefully supervised to ensure that, for example, the signs are put up correctly and facing the right way.

Access for passengers with disabilities is now a universal requirement. This requires level boarding and alighting of the vehicles, with a high level of safety for passengers on foot. This also benefits passengers in wheelchairs, as well as parents with children in prams and baby carriages. The elderly, who find steps difficult, will take kindly to such a user friendly system. For the operator, level boarding and alighting, especially with multi-door vehicle, means faster and safer boarding and alighting, typically 10–15 s, compared to over a minute with a single door bus. Ninety-five per cent of adults can manage a step of 200 mm (8 in) height but 100% can manage 50 mm, which is the usual margin of vertical tolerance between stop platforms and vehicle floor.

There are two approaches to achieve level boarding and alighting. One uses high platforms (900 – 1000 mm/36–40 in top of rail (TOR)) with high-floor vehicles (Figure 5.13). Increasingly, the norm is to use low-height platforms (typically 250–350 mm/10–14 in TOR) with low-floor vehicles (Figure 5.14). The option adopted may depend on local geography and politics. This is especially important if a significant part of the system is on narrow streets, where fitting high platforms would be a major challenge without disrupting the economic and social life of the city. A final consideration is that new low-height platforms will cost less than high platforms, such that for a given budget more

Figure 5.13 High platform light rail system, Manchester city centre

Source: Paul Steane

Figure 5.14 Nottingham town hall stop and low platforms

Source: Englandrover.com

low-platform stops can be built, further improving access to the light rail system. With stops close together, walking distances will be shorter and door to door journeys faster, making the system more attractive.

There are a range of options for constructing tram stops. Everything can be built in situ, or pre-fabricated with on-site erection. This latter makes sense when there is a large measure of standardisation of stop design. On the other hand, if every stop is different, then bespoke on site construction may be the most economic and practical solution.

Whatever the approach, erecting stops in busy city streets will create a disturbance. The design engineer can reduce this by minimising the occupation time with careful traffic management and maximising the use of pre-fabrication. If the end result is a beautiful addition to the street scene, then, like any other beautiful building, stops will soon be accepted as an integral part of the civic infrastructure. Beautiful and permanent stops will also be a reminder of an efficient transit system. Most residents will forgive the temporary disruption during construction for the long-term amenity provided. Visually attractive stations will help to build up loyal patronage, especially if maintenance and regular cleaning keep it that way. This can be enhanced by the use of planters, another maintenance requirement. Stops can also include photovoltaic (PV) panels on shelter roofs, providing power for stop lighting and signs.

As well as physically installing the stop, the supervising engineer will need first to check that the alignment of the tracks is correct. Alternatively, the stops can be built first and the track engineers fit the tracks to the stop. In either approach, the platforms will need to have straight edges so that the gap between vehicle and platform is uniform and safe, allowing the driver to supervise the opening and closing of doors. The relative alignment of platforms and tracks is crucial to the safe boarding of passengers and the efficient operation transit system. This may mean some minor redesign to fit in with the street and carriageway. For most new systems, the tracks come first.

Stops also need utilities – for example electricity to power lights and signs. Although in some locations, stop roof space could have PV panels and batteries to provide the power needed. In suburban locations, wind generators may also be installed at stops to provide power. There must be drainage and access to the covers of any utility plant located under platforms. Passenger information is an important part of the whole system design, with real-time departure times. Clear signage at every stop is vital, both for passengers boarding and alighting. Passengers need to be sure of the right stop and how to reach their final destination, on foot or by a connecting transit service.

During construction, a safe working site is the first sign of a new stop. Depending on traffic management arrangements, ideally the whole road, but at least one half of the road, will be closed to allow foundations to be laid if needed, and, if prefabricated units are being used, assembled on site. If the stop is particular to the site (Figure 5.15) then construction will have to be undertaken from the ground up, requiring a longer possession of the space.

Frontagers must have access to adjacent premises, and the site must have arrangements for the safe delivery and storage of building materials. Tram stop construction will normally only be during the working day (e.g. seven am to seven pm). This is particularly critical if stations are in residential areas. People do not like being disturbed at night. Site supervision and liaison with residents will be an important role for the design engineer.

Figure 5.15 Iconic light rail interchange stop in Strasbourg

Source: The author

5.2.4 Traffic management and priorities

A new light rail system should attract commuting and other journeys presently made by private car. It is logical, therefore, that the light rail vehicles should enjoy a measure of protection from traffic congestion, including priority at signalled road junctions. A typical LRV with 50 passengers has the same traffic impact as four private cars with an average of five people. In terms of increasing road capacity, trams are a very cost-effective approach. A twin track light rail with a tram every six minutes will have the same passenger capacity as a six-lane road with car traffic. The planning and engineering of traffic priority will need consultation and computer simulations to ensure that congestion is eased and that any spare capacity created is not filled by car traffic that is presently frustrated or suppressed.

During construction, there will be temporary traffic management measures to ensure the safety of construction teams and residents. Clearly, temporary traffic management will be implemented on a moving front, as track installation and stop construction progresses. Temporary traffic management could, however, become the final traffic management solution in some places. The main reason for this is that people are inherently conservative and do not like unnecessary changes.

No matter how comprehensive the publicity and advanced warnings before construction begins, experience shows that new traffic management measures, which may reduce motorists' route options and road capacity, are likely to result in negative publicity. This may have political impacts. The engineering team should therefore fully consult with residents and their political representatives, and aim for the optimum traffic management system. The fewer road network changes there are the better. Maintaining frontagers access to their premises during and after construction is usually a legal necessity, not just a courtesy. Discussing with frontagers their requirements for access is sensible consultation and a practical form of good manners.

The ultimate traffic management arrangements should be designed to ensure that light rail vehicles do not get stuck in traffic congestion. There are two main ways to achieve this. Where space permits, the light rail tracks can be physically separated from road traffic (Figure 5.16). The use of existing medians in dual carriageways or wide verges can be an economic and operationally beneficial solution (Figure 4.18).

Figure 5.16 Trams and buses sharing 'Trambaan' space in Gothenburg

Source: Google Streetview

Separation can also be achieved by displacing traffic from a lane and making it exclusive to tram (and in some places also buses). In Holland, such 'Trambaans' are also raised about 100 mm (4 in) above the rest of the carriageway, to reinforce this physical separation. Exclusive and paved tram lanes can of course be used by the emergency services and therefore help to speed arrival at incidents. This should ensure that the emergency services support priorities for light rail.

The second way to achieve separation at road junctions is by the use of time allocation. Typically, light rail operates at frequencies of about every 5 min, with a junction passing time of under 20 s. This means that for the other 4 min 40 s, road space can be used by other traffic, provided it is cleared when the next tram arrives. This 'green wave' can be achieved by trams announcing their approach and coordinated traffic signals linked into an area traffic network. This monitors road traffic flows and is triggered by approaching light rail vehicles. At junctions, there is a pre-empted green phase for a tram, cutting short other phases, which can be compensated in subsequent signal cycles by re-allocating the tram phases not needed. This keeps the main roads (and tram routes) flowing and confines queues and congestion to the side roads.

Probably the most advanced traffic priority system was pioneered in Zurich (Figure 5.17), where traffic lights automatically give a green signal to an approaching tram (or bus). Contrary to popular belief this actually speeds all traffic, since trams and buses not stopping at the junction spend less time clearing the junction. This releases capacity for other traffic. While the Zurich system speeds all traffic, it speeds public transport significantly more. This has resulted in a modal switch from car to tram (and bus). This is a win-win situation and in line with the prediction of Mogridge (1993)

How can this level of junction priority be achieved? One approach is to programme a tram green phase for every signal cycle that coincides with the timetabled crossing of a tram. If tram services run 100% reliably, this would be the simplest and cheapest option. Experience shows that 100% operational reliability is not always possible. Zurich and other European cities use tram (and bus) detectors so that an approaching tram will immediately call up a green phase in the signal cycle. Once passed there will be extra time to compensate the other phases and traffic, during the signal cycles when there is no tram.

Figure 5.17 Tram priority in Zurich. Notice the special signals

Source: SPG Media Group Ltd

Such detectors can be simple magnetic loops in the road, tuned for long vehicles, so will not be triggered by cars. This is the system that has worked for more than 40 years on the Runcorn Busway, giving buses running at 65 km/h (40 mph) uninterrupted crossing of road junctions. More usually, a beacon on the tram transmits an identification signal, both to trigger a green phase and, where there is a tramway junction, to change the points and call up a turn filter phase so that the manoeuvre can be undertaken safely without stopping. The ability to identify individual vehicles can also be used in timetable regulation and service management. Trams running early could be made to wait for a green, whereas trams on time or running late will always get an immediate green.

The mechanics of traffic management during construction will be different from those required after the opening of the tram service. During construction, temporary and moveable barriers and signs will be used. These will prevent vehicles or pedestrians tripping or falling into any excavations. They will also direct traffic to alternative routes, enabling the construction site workers to be safe from traffic.

Permanent traffic management arrangements will require the erection of signs, normally during the construction phase but covered over. This means that when the new system is ready, the signage can be unveiled as appropriate. Traffic signal co-ordination will need more careful preparation and may also have to be fine-tuned when opened. Experience in major city-wide traffic management shows that drivers take about two weeks to adjust to new traffic signal settings and 'green wave' speeds (Matsoukis, 1980). Historically, network traffic signal plans were calculated off-line, working on a fixed cycle time, although with phase splits changing throughout the day to reflect varying traffic flows.

Control computers are now able to optimise a traffic signalled network in real time based on actual traffic flows detected. In practice, because traffic flows are nearly predictable on a daily and weekly basis, sub optimisation with local detection and control can be used. This makes giving priority to light rail vehicles at road intersections easy. While there is considerable international standardisation of road traffic signals, with red, yellow and green lights arranged vertically with red at the top, there is no similar international standards for light rail signals in streets. These must therefore reflect local operating practice.

5.2.5 The vehicles

While trams are normally made remotely from the city construction sites, the commissioning engineer will need to monitor progress, both to ensure that the vehicles ordered meet the required specification and are delivered on time for the system opening. There will be a period when drivers are trained with the new vehicles, to become used to the method of operation and the operating characteristics. Similarly, the maintenance team will be trained by the manufacturer in keeping the fleet running. The car builders will be required with the operator to test the vehicles and subsystems as work progresses. The commissioning engineer should be present when these tests are undertaken. This will allow confirmation that the vehicle operates as expected and that recovery from emergencies and breakdowns is fully understood and documented.

The car builder will be required to provide

(*a*) a maintenance manual
(*b*) driver's/operation manual.

If the car design is a variation of an existing vehicle, then the documentation provided should be straightforward. On the other hand, if the car is bespoke for the new system, then there may be more problems with the after service.

Some new car orders include a period when the builder is on site and undertakes maintenance for a set period. This must include training the local staff to take over after this period. There will also be written into the supply contract a period when the car builder will be required to correct any defects that occur under warranty. Alternatively, the car builder may have a period when the fleet is maintained by them on an ongoing basis.

This was a problem for Siemens over the low-floor Combino, first supplied to Potsdam in 1998, which after about 100 000 km developed fatigue fractures in the aluminium body. By this time, some 300 had been supplied worldwide. Siemens had to introduce retro-repairs, costing about €100 000 per vehicle. Other low-floor trams also have had problems not shown up in the usual testing. Those with fixed trucks create further problems relating to rapid tyre and rail wear. They also have wheel squeal noise problems transiting small radius curves.

5.3. Commissioning

Before a new light rail system or line can be opened for passenger traffic, there will be a period of commissioning, training and approval for operation. There may also be a period without fares being charged to allow passengers to gain experience of using the system. A free trial is common practice for commercial products and, if nothing else, is good public relations for the system promoter.

The first stage of commissioning is the static testing of all equipment. Other commissioning will almost certainly be undertaken in parallel, rather than sequentially, unless there are any secondary relationships. Additionally, there will be a series of emergency practices, involving both the operations team and the emergency services. This provides important experience, since the history of 'crashes' (as happened in Oslo in October 2024, shown in Figure 5.18) shows that the control team cannot image all possible solutions and needs to have recovery plans available. The rail licencing authority will also expect this.

5.3.1 Electrical supply and OHL

Commissioning the power supply for the OHL and the operation of substation switching will ensure that a safe supply of power is available. Tests will include leakage currents, response times to switch on and off commands, the ability to switch out substations, while continuing to supply

Figure 5.18 Tram derailment in Oslo, October 2024

power from adjacent substations. At this stage, emergency procedures can be practiced in response to short circuits, vehicle crashes and other events where the OHL might present a danger – for example, a crane snagging and pulling down the OHL. Switching off power automatically in such events should be accomplished in milliseconds.

Track continuity between one end of the system and the other is vital to identify any broken rail joints, or poor return current connections to substations. These can create stray currents in an earthed light rail system. As the stray currents involved are normally in the milli-amp range, ultra-sensitive test equipment is needed. Repairing any broken rails before service starts must be a high priority, because even if the other rails can provide a circuit to the sub-station, the risk of a break in the other rail would be problematic.

Similarly, the insulation of the traction poles and street furniture is important. Voltages over 100 V are normally fatal, so the OHL at about 750 V must be carefully tested for insulation. This is especially needed in wet weather. Rain can provide current paths, including making traction poles or other metal street furniture live.

Normally, light rail vehicles should be delivered fully tested. For completeness and as part of the static testing, vehicles can be further tested in the depot. The first test is to ensure that under all operating conditions there is never a short circuit within the vehicles. Should a fault occur, then the system within the LRV must automatically shut down. Failing that, the local substation must immediately switch off the power to the OHL. A second test is to ensure that current return is satisfactory. Should the tracks be covered by insulating material (e.g. sand on tracks as shown in Figure 5.19) or the wheel current return bushes fail, then the vehicle body must not become live, endangering staff and passengers.

5.3.2 Trackage

Tracks in streets effect two classes of users, the light rail vehicles and other road users going along or across tracks. For other road vehicles, clear lane markings and signs are the best way to ensure compatible interactions, assuming that contractors have laid tracks compliant with the road surface, particularly important for the safely of pedestrians to avoid trips when crossing the tracks.

Figure 5.19 Sand blown across tracks in Blackpool

Source: The author

Gauging is the first stage of track commissioning. This ensures that tracks are compliant with the gauge tolerance. The second stage will be the geometry both along the tracks (alignment) and across the tracks (level), both of which will affect the safe speed of operation and the comfort of passengers, especially those standing. The safe ergonomic forces that passengers can accommodate are well known. Comfortable accelerations and jerk rates are also known. The curvature of tracks and the planned operating speed of vehicles will determine safe and comfortable lateral accelerations (Equation 5.1).

$$F = M V^2/R \tag{5.1}$$

where F = the centrifugal acceleration force,

V = operating speed and R = radius of the curve.

Where the maximum centrifugal comfortable acceleration = 1 m/s² (*c.* 3 ft/s²), the relationship between curve radius and maximum speed can be plotted (Figure 5.20).

Commissioning must include public education, especially when there has been no recent rail operation on roads. For cyclists this is important, to avoid accidents due to bike wheels getting stuck in rail grooves, the cause of an early fatality on the then newly opened Sheffield Supertramway in 1996.

For light rail vehicles, the correct operation of points (turnouts) and signals is crucial to ensure safe and efficient operation. This will be demonstrated by a vehicle operating at low speed, and driver training to ensure that the right route codes have been used. Should a malfunction occur – for

Figure 5.20 Comfortable curve speed at 1m/s² lateral acceleration

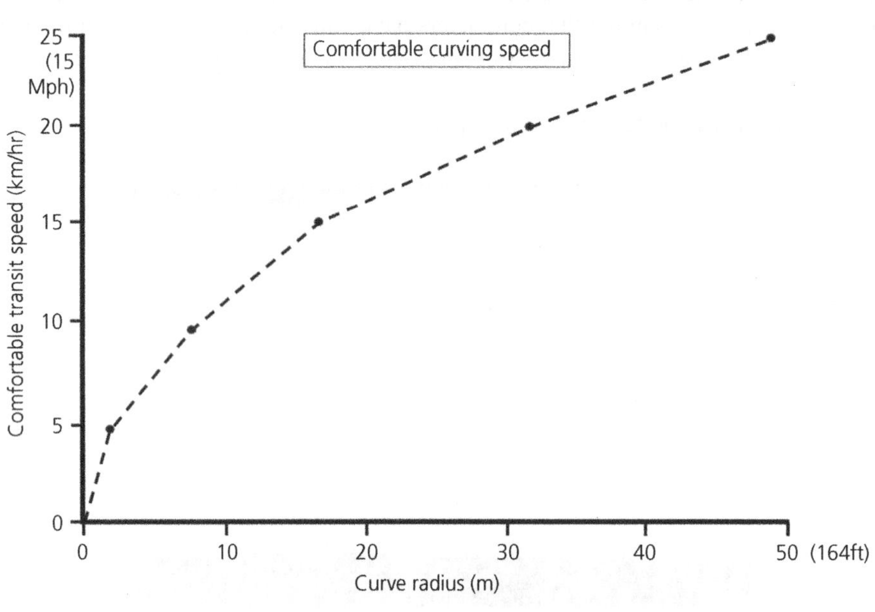

Source: The author

example, wrong point setting – the driver has time to stop before damage is done to the vehicle, track, road vehicles or roadside furniture.

5.3.3 Stops

The team that designed and builds the stations should be the first to commission, to ensure that all systems are working as specified and that the structure has been built as designed. At this stage any snagging works can be identified. The first place that passengers, in practice, experience a light rail system is at a stop, while waiting for a departure. Safe and easy access to the station, including crossing any road carriageway to reach the platform, should be the designer's aim. Checking that this is the case could use 'mystery shopper' volunteers who normally undertake consumer testing. They will be neutral, professional and help to identify any problems not picked up before. These can include the legibility of signs at different times of the day and the evenness and grip of the platform surface under different weather conditions.

Similarly, groups of people – for example, pensioners – can be recruited to board and alight from a light rail vehicle at the stop. This will allow the flow of people on the station platform to be monitored. It will also allow emergency evacuation procedures to be tested both for light rail vehicles and the whole stop platform area. Pinch points can be identified and measures to ease these installed.

Most importantly, stop commissioning should also include orientation and direction signs for passengers to navigate the system. It may be prudent to build one of the stations early so that there can be consumer feedback, rather than having to retrofit modifications later to all stations. All remedial works identified in this process will need to be programmed before the formal opening of service.

5.3.4 Vehicles

After the stops, the second most important part of a light rail journey for passengers is the ride in the LRV. In spite of a lot of preplanning, mock ups and ergonomic studies, trying out the light rail system with real passengers will show up areas where small adjustments might make a big difference to operations – for example, emergency signs. Similarly, changes in operational management might make an important difference in the way passengers use and enjoy, rather than endure, the service. Part of the commissioning will include emergency evacuation from simulated accidents. This will also involve the emergency services.

While a lot of minor accidents occur at light rail stops – for example, passengers tripping – major accidents are usually away from stations, often in junction areas. This has certainly been the experience in Manchester, Sheffield and Croydon, where road vehicles disobeying traffic signals and collided with light rail vehicles undertaking their priority use of the junction. In the most serious case, a heavy truck crashed into the side of a Manchester light rail vehicle, derailing and pushing it sideways. A police car was also badly damaged, in San Diego, when turning left across the light rail tracks into the side of a light rail vehicle going in the same direction.

For an emergency to be properly tested, the tram should have a range of passengers, including teenagers, babies and small children with parents, adults, pensioners and people with disabilities, including those wheelchair bound. Co-ordination between light rail and emergency service staff will be important for the safe evacuation and medical treatment of injured passengers or operators. From experience in other rail systems, emergencies cannot be managed in an ad hoc way. The training of staff in the safe and courteous handling of passengers is an important foundation for a safe and successful long-term operations.

Thankfully, light rail crashes are rare but the consequences can be severe. In 2010 there were only two light rail crashes involving injuries worldwide: Melbourne (11) and Rotterdam (35) (Figure 5.21). Being so rare, these, however, make headline news around the world. Nevertheless, having staff fully trained can firstly avoid accidents and secondly mitigate the effects when accidents occur. Sadly, since then in 2016 there was a seven-person fatality, with 62 passengers injured on the Tramlink system in Croydon, due to entering a curve too fast, derailing and toppling over (Figure 5.22).

5.3.5 Depot

The depot serves several functions, including daily cleaning and maintenance, overnight stabling, routine maintenance, accident repairs, staff daily briefings, training and hygiene, system control, and security. With the infrastructure fixed, commissioning the depot will only show how the fixed assets can be managed better or more safely, and the staff trained efficiently to operate the system. Over time, staff may find short cuts and attempt to abuse facilities. The role of commissioning is to identify how these might arise and put into place procedures to ensure that staff perform their functions safely.

Figure 5.21 Crash in Rotterdam, March 2009

Source: Leo Roubos

Figure 5.22 Croydon fatal tram crash 2016

Source: RAIB

Commissioning the depot will include testing all the equipment, including hoists and jacks for lifting light rail vehicles. If the control centre is in the depot, all its functions, including alarms, switches, communications equipment, radio and so on, will need to be proven. Emergency exercises will be co-ordinated from the control centre. Liaison with the emergency services will include inspection tours and meeting the control centre team. Safety features like drains, ventilation and ducting will also need to be commissioned. Finally, all health and safety signage and facilities need to be checked and the depot staff fully trained.

5.3.6 Service

In commissioning the system there will be a period of running the service without passengers, so that operating staff can become familiar with the equipment and its operating parameters, as well as the route and tracks. This period will be between a week and a month. This allows service timetables to be fine-tuned, especially for any peak/off peak adjustments. For journeys in peak times, with higher volumes of passengers, boarding and alighting will take longer and stop times will increase. Therefore, there might be effectively two timetables, one for the peak and the other off peak. This is especially the case where not all stops are compulsory, and will avoid holding up a following light rail vehicle, or missing its allotted time at a junction. For most stops, a dwell time will be between 10 and 20 s, allowing for doors to open and close and for alighting and boarding passengers. With level boarding/alighting this is faster, typically on average about 2 s per passenger. With wide doors, two passengers can alight simultaneously and with several doors – for example, with four doors eight passengers can alight in 2 s. Heavily used and interchange stops may need individual stop times. On most LRT systems, however, there will be little daily variation of trip times.

System commissioning will also include two-way radio, which will allow thc opcration to be managed in real time, should problems arise. It also allows staff rostering, including crew relief for meal breaks and the end of shift, to be monitored. Slick change overs are important for running a reliable service. A terminal station is the best place for staff to change shifts, since the vehicles will be timetabled to stay there for a few minutes and any delay in staff changeover will have a minimal effect on service reliability.

When the Sheffield Supertramway was being commissioned during November 1995, a sudden snow storm halted all other traffic in the city, except the trams running empty. That certainly left an important message about the weather reliability of trams but also frustrated travellers stuck in snow bound cars, who would have been happy to take a tram.

5.3.7 Out of service and emergencies

No light rail system can expect to operate without experiencing occasional out of service problems, like rail vehicles breaking down or a road vehicle blocking the tracks. Less frequently, there will be emergencies, some of which will be caused on the tracks – for example, a light rail vehicle or other road vehicle crash. Some emergencies will be off tracks, like a fire in a building adjacent to the tracks. The fire service will require the OHL power to be turned off to allow fire fighters safe access to and the use of water at the scene. This will also mean that the tram service has to be suspended. Fortunately, such emergencies are rare and, with radio links, the control centre can advise passengers at stops and on trams about any delay.

Experience has shown that it is not realistic to expect the control team to ad lib its way out of every situation. The reason for this is that in normal service, the amount of information handled by the control room staff is minimal (Cuffe, 1989). During an incident, the data received by the control staff increases many fold. Often the wrong information is acted upon, making the incident worse. Part of the commissioning period should be the training of the control team to handle incidents, both through training exercises and as in Section 4.9.4 in undertaking simulated incidents with real people, on site – for example, at a stop or road junction. Emergency plans can then be prepared, reducing the need for inventing solutions.

An emergency evacuation exercise of a City Class tram, with 180 passengers was undertaken in Birkenhead, filmed and timed. The evacuation took 35 seconds. Half the passengers used just one of the four available double width doors. The other half were spread between the remaining three doors. Had the 180 passengers used all doors evenly, then the full evacuation could have been completed in under 24 seconds. This demonstrates that in emergencies people do not always behave rationally. Nevertheless, the extra 11 seconds represents nearly 50% longer than might have been the case. This kind of behaviour must be included in emergency plans.

This allows the performance of the control team to be fed back in a constructive way, so that, come a real incident, the team will be ready to take the decisions that best liaise with the emergency services and maximise the safety of passengers on the system and others nearby. Strategies will be developed for restoring normal service quickly. Such training exercises should be repeated regularly to maintain efficiency.

5.4. Maintenance and repairs

5.4.1 Track

The two most common track problems are wear and corrugations. Track wear will be most noticeable around sharp curves (under 30 m radius) (Figure 4.16), where there will be flange grinding into the rail side of the outer rail and into the groove check rail on the inner. This can lead to the loss of the check rail, especially if also impacted by heavy road vehicles. Flange lubrication is a way to reduce this wear, as well as the wheel squeal which is associated with it. Similarly reducing vehicle operating speed around curves will also reduce this wear.

Initially, rail grinding will restore the rail head profile and therefore reduce the risk of derailments from the flanges of the wheels climbing worn rail sides. Ultimately, even reprofiling cannot work, since the track will be out of gauge tolerance and the groove check rail will become worn through or broken off (Figure 5.23), creating another derailment source. This premature wear of curved check rail led to the early replacement of a curve in 2007 on the Croydon tramway near East Croydon Station.

Rail head corrugations are worse when rails are laid directly on rigid bearers, whether concrete pads or sleepers. Corrugations are formed by a fatigue mechanism creating a series of hard ridges across the rail head, with a regular wave length. Tram rail corrugations usually have a wavelength of about 100 mm or less. These are due to hunting of the wheel sets across the track. This work hardens into zones about 2 mm wide and deep and allows the wheels to abrade the softer head between hard spots.

Unless treated early, the hard zones will propagate vertically right through the rail head. To prevent that, the only remedial treatment is by rail head grinding to below the zone of hardness, usually more than 2 mm depth. Rail grinding will not remove that problem, since it is a result of tram wheels interaction. Ultimately rail replacement will be needed regularly. This has a high cost, since otherwise serviceable rails have to be scrapped prematurely and possibly the tram service disrupted.

Figure 5.23 Croydon broken check rail

Source: The author

Failures of crossings and supports are a sign either of poor design or skimped maintenance. The example from Sacramento (Figure 5.24) shows a railroad design crossing, used on the light rail system. The paving blocks are in a poor state and show that LRV wheels have been over running. The most likely cause of this failure is due to wheels dropping into the crossing gap, hammering and damaging the foundation and disturbing the paving blocks. On a railroad this is not a problem, as thc crossing will be supported on an 'elastic' tie (sleeper) and there is no paving.

For segregated tracks, monitoring the ballast condition and drainage will be important to maintaining a safe and comfortable ride by LRVs. Where grass has been sown around the tracks, the

Figure 5.24 Crossing failure in Sacramento

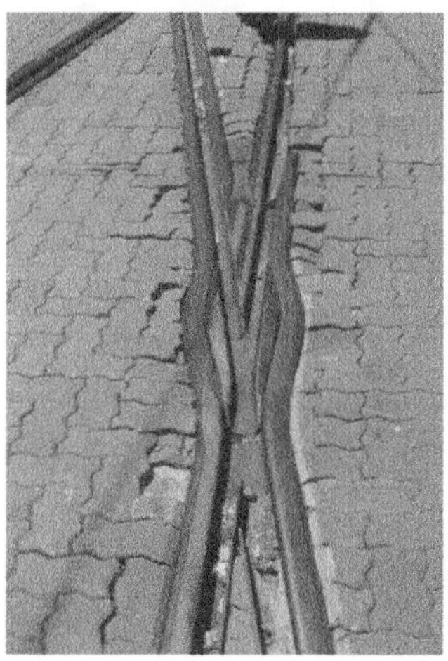

Source: The author

passage of LRVs will keep the grass mown but outside the LRV swept path, grass will need to be mown regularly during the growing season. Alternately green slow growing plants with a lower maintenance could be used – for example, sedums.

Regular maintenance tasks for tram tracks include keeping the grooves and drains clear of debris, so that in rain the rail head remains above any standing water. The lubrication of switches is important, as is keeping tongues and blades clean of debris, so that they operate smoothly, without danger of trams being derailed in not following the right path. Normally, switches are inspected weekly, unless a tram driver reports a problem.

Where tracks are embedded in streets, debris is also deposited by others – for example, motor vehicles, pedestrians and so on. Track grooves can be cleaned out mechanically – for example, with large vacuum cleaners. Switches require manual attention, at busy places weekly and elsewhere at least monthly. In frosty weather (Figure 5.25), where tram tracks do not share salted roads, the operator will have to plough or sweep snow from the tracks and defrost switch blade pivots and blade-throw slots. Some systems also operate LRVs overnight to keep tracks clear for the start of passenger service in the morning.

The operator will also have to observe the state of the highway pavement around tram tracks, since any delamination between the highway pavement and rails will lead to water penetrating the track foundation. If shared with heavy road vehicles, water pumped into cracks by road vehicle wheels can lead to track failure, as has happened several times with the shared tracks in Moseley Street in Manchester.

The delamination of road pavement from tram rails can be caused by both road vehicles and railcars but the light rail system operator is the one who will suffer the damage and therefore has an incentive to rectify this as quickly as possible. Most track maintenance and many repairs can be

Figure 5.25 Snow blocked light rail track

Source: The author

undertaken without having to suspend services, although road traffic may need to be temporarily diverted from the work site. Special maintenance vehicles equipped with power tools have made track maintenance much faster, less costly and safer.

5.4.2 OHL and power supply

Overhead lines with carbon contact strip pantographs should last about 30 years before wear reduces the mechanical strength or increases the electrical resistance to a point where replacement is needed. Normally, OHL properly installed should be nearly maintenance free, but will need repairing if damaged by a broken pantograph or hit by a high road vehicle – for example, a crane boom on a transporter. In this case, repair to the original line profile is needed, depending on the degree of damage.

Traditionally, a rail tower wagon has been used to gain access to the OHL. Today, a standard road self-propelled mobile (Cherry Picker) work platform is the most economical, as well as giving the most flexible approach. Such a vehicle can be fitted with flanged dolly wheels for use on unpaved or off-street tracks. The maintenance or repair of OHL can also be required by adverse weather conditions, wind damage or thermal displacement. Similarly, if the OHL is suspended from adjacent buildings, demolition will require the planting of a (temporary) pole or a new fixture to an adjacent building to support the OHL.

Regular checking is needed of the OHL power supply cables to ensure that the insulation has not deteriorated, joints failed or there has been some mechanical damage. Similarly, the feeder box isolators and circuit breakers in the sub stations need regular (monthly) checking to ensure that they are working in both ‘off’ and ‘on’ directions. In particular, checking must ensure that there are no short circuits (to earth).

5.4.3 Vehicles

The manufacturers of tramcars should supply a maintenance manual and schedule. Two of the biggest areas for maintenance or repair are the wheel (profiles) and passenger use damage.

Light rail wheel profiles can be damaged by grinding rails sideways into flanges, from wheels locking during braking and wearing a flat, or from spinning on acceleration and therefore damaging the whole tyre metallurgy by overheating. The normal method of recovering damaged wheels is using a lathe. Reprofiling the tyre (and or flange) is achieved by turning back to the required profile, if there is sufficient metal.

Some systems will have underfloor lathes in the depot and can turn a pair of wheels on the same axle in situ at the same time. In others, the whole axle/wheel set has to be removed and put into a bench lathe. Finally, there are operators of vehicles (e.g. PCC, City Class and so on) where the wheels can be individually removed and re-turned on a bench lathe (Figure 5.26).

Turning the wheels will remove metal from the tyre and shorten its life. Reducing wheel damage in normal operation is possible using flange lubricators, which will also reduce rail damage on curves, and having slip/slide (spin/skid) protection in the traction control system. This prevents wheels locking during braking and, in acceleration, keeps the wheels to the maximum track adhesion available, without spinning/slipping. Keeping the wheel profile true not only reduces rail maintenance and, by less vibration, reduces damage to the vehicle but also improves the ride quality.

Figure 5.26 Removable wheels: left – City Class; right – PCC

Source: The author

Figure 5.27 Door glass scratched and graffiti in depot

Source: The author

Passengers can damage rail vehicles accidentally and also, unfortunately, by vandalism. Dirty clothes, seat abrasion and shoe damage to floor surfaces are the most common regular maintenance tasks. Keeping light rail vehicles clean and looking smart is no different to hotels, hospitals or any other facility with public access. There is evidence which shows that vehicles kept clean will not be spoilt so much by passengers. Indeed, operators can enlist passenger support in keeping the system presentable, through social pressures – for example, by them not leaving litter.

Vandalism, on the other hand, is more difficult and the usual remedy of making surfaces more vandal resistant will usually deflect attention to other areas. Window scratching (Figure 5.27) and graffiti are two areas of vandalism that are expensive to cure and have the biggest impact on regular passengers. In the extreme, glass replacement may be needed, taking the vehicle out of service, which is expensive. For shallow scratches, acrylic based fillers are an option.

The installation of CCTV on vehicles can provide evidence to prosecute vandals. It is more positive and cheaper to deter vandalism to avoid the need to repair or replace. Here, applied psychology can be effective, including supervision by staff and other passengers, and programmes with schools and youth groups to channel these destructive energies into more creative avenues.

More recently, with the near universal adoption of smart phones and similar devices, operators have found that providing free WIFI significantly reduces vandalism and anti-social behaviour on vehicles, as passengers are entertained by their phones, or busy texting friends during the journey. Wi-Fi on one route was provided as a commercial experiment where, almost immediately, vandalism stopped. The cost of the Wi-Fi was more than covered by not having vehicles out of service, while vandal damage was repaired. On some systems, classical music played has been shown as a way to stop the congregation of young people, who can be un-nerving for some passengers. Music can also be used to create a calm riding atmosphere.

5.4.4 Stops and other infrastructure

Light rail stops need regular cleaning. Where trams run in street, the local municipality can clean platforms at the same time as the rest of the street. For off-street lines, dedicated cleaning teams are needed, on a daily basis to collect litter and to repair damage. Smashing shelter glass at secluded stops can be a temptation for vandals. Stations designed on the principles of defensible space are less likely to be vandalised, especially if social control measures are adopted.

A rapid repair strategy can also be effective in reducing such damage. A demonstration of this was in Bradford, where the replacement of broken bus shelter glass was shortened from within 1 week to within 24 h, and this virtually stopped shelter damage within 1 month. British Telecom had almost the same experience with telephone kiosk damage.

The lesson from these experiences is that pre-emptive maintenance is the best strategy to avoid expensive repairs. Having a pleasant waiting environment is the best way to build passenger confidence in the system and, with that, patronage and revenue to fund the service. A well-used tram stop also has the benefit of social pressure reducing or avoiding vandalism or graffiti. CCTV cannot replace the presence of people but can provide evidence if needed for prosecution, if the system is working and the cameras are pointing in the right direction.

5.5. Refurbishment and enhancement

All light rail systems require regular updating and refurbishment, both to replace physically worn out assets, as well as providing a more contemporary aspect for stops and vehicles. This must be included in the business plan. It can be funded by regular depreciation payments from the annual accounts, or by having a year with an operating deficit, which can be offset over the following years, against profits from operating surpluses. Whichever way these are funded, if they are not in the budget, then the system cannot maintain long term good visibility and hence viability.

Refurbishment, however it is funded, needs to be carefully planned and involve existing passengers and target passengers, to get styles and colours that will win the greatest support. Refurbishment must have engineering inputs as well as design and styling inputs. The replacement of equipment will usually be based on a financial appraisal – for example, will a new item be overall less expensive than continuing to maintain the old item? This evaluation will also include improved performance and the ability to satisfy passenger expectations. With equipment development, it is likely that the replacement equipment will have more functionality and be less costly to operate and maintain.

5.5.1 Vehicles

Passengers often compare public transport vehicles and light rail cars with contemporary private motor cars. Railcars can never have every gizmo found in a modern motor car. In terms of presentation, style, comfort and convenience, however, LRVs need to look as good, and with appropriate industrial designers, there is no reason why light rail vehicles cannot have a superior interior presentation to automobiles. A parallel comparison is tram stops and hotels.

Another reason for refurbishment is changes of operating legislation, requiring further comfort, or safety facilities for passengers or staff – for example, the requirement to make transit vehicles accessible for people with disabilities. Retrofitting facilities usually involves compromises, or reducing the comfort or capacity for other passengers.

Similarly, the need to provide real time information for passengers inside the rail vehicle, as it progresses along the route, is not just a legal requirement in many countries but an important marketing tool. This is especially the case where attracting car commuters is critical, since passenger orientation is important to building passenger confidence. Car commuters are used to well-signed roads to get to their destination. For transit, the signage must be at least as good and the operations as reliable as clockwork.

Probably the easiest and least expensive way to change the look of a light rail vehicle is by a new colour scheme (Figure 5.28) and new seat coverings. Where seat cushions are well worn, they should be replaced; as a minimum the cushions, but better still new seats. This will allow a new seating arrangement and will be a further enhancement, both for the looks of the vehicle and for the comfort of passengers.

Figure 5.28 Supertrams before and after upgrading

Sources: The author and Stagecoach plc

During the operating life of a tram fleet, a record of reasons for delays and breakdowns will be kept. The most frequent cause of such problems will be known, most probably due to vehicle door failures. During refurbishment, improvements to these critical parts can be implemented that will reduce the down time in future operating periods and improve the fleet reliability and therefore passenger satisfaction.

Rail vehicles can have a life of about 30 years. In practice this can be extended to 40 years, depending on their operational reliability. The finances of the operator may make half-life refurbishment a sensible option. The rolling stock refurbishment industry has developed considerably in recent years. Refurbishment can also include operational life extension. For passengers and staff, a refurbished vehicle can seem new, and refurbishment costs a fraction of buying new vehicles. The decision to refurbish will consider the availability and cost of spare parts, energy consumption and feedback from passengers.

Clearly, such makeovers will require vehicles to be out of service. This has two consequences. Firstly, the fleet will be reduced for normal service, which may result in a less frequent service. Secondly, workshop space will be occupied, which is then not available for routine maintenance. An alternative to the second is to subcontract the refurbishment to an outside company. This means that a rota of vehicles will leave the system. Which of the two is chosen depends on the level of refurbishment and how accurately the programme can be scheduled.

The operator may, however, acquire some spare vehicles, by purchase or loan, to cover for those taken out of service for refurbishment. This is usually possible where neighbouring cities have similar or compatible vehicles, that can inter operate.

5.5.2 Stops

Refurbishing stops may be driven by changes in legislation, as discussed for vehicles, especially the need to make systems accessible for people with disabilities. Level or ramped access can usually be arranged for low-platform light-rail stops in streets, but may have a down side in that elderly passengers, especially in wet or frosty weather can slip and fall on slopes, a common cause of fractured hips or legs. Providing a handrail on each ramp is one way to increase passenger safety.

On the other hand, making stops and vehicles accessible also helps parents with young children in buggies or prams, for whom the only alternative travel might be the family car, when available. Indeed, it was the unexpected off-peak use of the Manchester Metrolink, mostly by women with small children and from one-car households, that enabled the first line to reach early operating profitability. It also allowed a higher frequency to operate throughout the day for a turn up and go service. This generated further trips in other sectors of the population.

Closing stops for refurbishment will also be challenging, since the busiest stations will be the ones which will have the most normal wear and tear (Figure 5.29). Undertaking works at these stops while keeping in use needs a critical programme, including overnight work when no passenger service is operated. As with vehicles, changing the décor of stations is the fastest way to improve the passenger waiting environment. The refurbishment can also include better information and a public address system. Other additions like planting and vegetation (Figure 5.30) can increase the need for care and maintenance, as flowers die and to avoid planters becoming litter bins. On the other hand, 'greenery' makes a place more attractive.

Figure 5.29 Milan street stop – candidate for refurbishment?

Source: The author

Figure 5.30 'Green' tram stops

Anecdotally, young people meeting and hanging around tram stations can be disconcerting for many passengers. The Tyne and Wear Metro addressed this by using the public address system to broadcast classical music, which had the effect of moving groups of young people away and perhaps making some passengers who like classical music happier.

Where rail stops are in street, any refurbishment must also consider the local environment of street furniture, the needs of building frontagers and so on. It can also provide an opportunity of allowing a themed refurbishment sponsored by local businesses.

A light rail operator will try to run a frequent and reliable service, so that passenger waiting time is kept short. Nevertheless, shelter will be important to protect users from inclement weather, no matter how short the wait. How much shelter will depend on passenger numbers and service frequency. What sort of shelter will depend on budget, house style, design innovation and the prevailing weather conditions.

There are plenty of box like 'bus shelters' on the market, that can be bought 'off the shelf'. A progressive operator will, however, be trying to offer an 'up market' image and therefore will wish to provide shelters designed with flair. This will give the rail service a unique identity, which is important in creating and enhancing the brand image. A light rail system should be iconic in its city. At a time when multinational brands festoon city streets worldwide, a light rail system can be one, if not the most, important local civic identifier, for which stops are a major contributor.

Good transit design creates a loyal patronage base. The 1930s schematic London Underground Map is now copied throughout the world, because it simplifies the perception of transit connections, making it understandable for passengers who cannot read a geographical map. Building a local identity is important for a strong brand locally. The design of stations is central to this.

5.5.3 Other

Other aspects of a light rail system that benefit from regular maintenance or updating include the traction poles, equipment boxes in public places and the public face of the depot. Apart for the stops, traction poles and overhead power lines are always in the public eye. Unless a motor vehicle collides with a traction pole, there are normally few reasons to pay attention to this aspect of the system. Regular painting can make a big impact on the urban streetscape, and, if integrated with the stop house style, is an important marketing tool to building and retaining patronage of the tramway. Good design and the choice of livery are not expensive, when regular maintenance includes painting.

Where climate permits, hanging baskets of flowers can be put on traction poles to improve the city ambiance. In some systems, vegetation is allowed or encouraged to grow up poles. This can provide seasonal variations, as well as homes for local fauna. Local residents can be recruited to water planters regularly. Similarly, traction poles can be used to fly flags and banners as part of a local carnival or festival. While these are not directly required for light rail operation, they do help to establish the operator as part of the local community, a friend of the environment and, therefore, within the collective consciousness.

Depots are seen mostly by those who live or work nearby. It is important that depots are good neighbours, presenting a tidy and clean image (Figure 5.31). Like any other engineering facility,

Figure 5.31 Tidy depot in Berlin

obsolete equipment can accumulate, so regular tidying up and clearing out will not only help maintain a good image but also improve working efficiency. Better staff morale can be fostered by a policy of good housekeeping in the depot. Working in an organised and tidy facility improves staff productivity and reduces the risks of accidents.

5.6. System extension and development

Very rarely can a complete light rail system be built all at once, if for no other reason than the availability of funds for construction and equipping. Building a system in stages is therefore the norm. This also means that patronage and, therefore, revenue will increase as each stage is opened. More lines generate a synergy of increasing travel opportunities, patronage and revenue. Having revenue coming in early gives funders confidence for further investments. Confirming the projected figures for the patronage of an extension is critical in all business plans, guaranteeing that future investments will be financially viable. This was an area of criticism in the UK National Audit Office Report of April 2004 into light rail projects. Refining patronage forecasts from actual usage means that the economic evaluation of future lines can be made with greater confidence than for the initial line.

In building the first line, there will be obvious places where future route extensions can begin. The question for the design engineer is whether the junctions should be laid in anticipation of future extensions, or whether the operating line should be disrupted when future extensions come to be built. In the urban centre, usually it makes sense to install junctions in anticipation of extensions, whereas in the suburbs, the disruption of installing new junctions will be less, so can be left until needed. In any case, even over a ten-year period, it is not unusual that, when considering the case for a full network, priorities could change and new urban developments may justify construction of a line in advance of established areas, especially if the developer is willing to contribute to the cost.

5.6.1 Increase route capacity

Before new lines are built, it often makes considerable sense to see how much extra traffic can be attracted to the existing line(s) by other means. If this can be done with operational, marketing or management measures that need only limited capital investments, all the better. There are three ways to increase capacity within the existing capital and infrastructure assets. These include increasing the capacity of rail vehicles, or increasing the number of vehicles per hour.

The first can often be achieved by replacing seats with standing space. This adds more effective space in peak periods. Replacing four seats with standing area will provide about 15 spaces, a net increase of 11, which in a 200-passenger vehicle is about 6%. A compromise is the use of tip-up seats, which at quiet times can be sat on, whereas at peak times, the equivalent standing space will be available. Passenger surveys can determine the most acceptable way in which extra capacity can be achieved. It can also have a dual use – for example, in providing more space for wheelchair or pram space.

The second way capacity can be increased is by lengthening articulated cars or coupling units into a consist/train (Figure 5.32). This adds capacity without any change to the timetable or service frequency, and without additional staff costs. Many new LRT systems are designed for two car operation and thus no changes in stops are needed. The power supply may, however, need to the strengthened as two coupled cars will draw more power on starting than a single unit.

This method increases capacity by having more vehicles per hour formed into trains. Many light rail systems use two or three coupled cars. Running longer trains of course means longer platforms

Figure 5.32 Long platforms for Dusseldorf light rail two-car consists

Source: City Transport Information

at stops. Traffic management and signal settings may need to be modified to allow the safe crossing of busy junctions. Both these will require investment but less than building a new line and might achieve a similar patronage increase.

Alternatively, more trams per hour can be scheduled. If the vehicle fleet is fixed, then either the running times have to be reduced, or the service recast so that there is a more intense service over the busiest part of the route. With operational experience, drivers should become more competent and confident. Running times, frequencies and train length should be constantly reviewed and revised as needed. When the timetable can be tightened up, it will improve system efficiency. If more runs per day are achieved, this increases both vehicle and staff productivity. A fuller discussion on this can be found in Section 5.6.4 below. There is always scope for improving performance, and every extra journey squeezed from the existing fleet reduces unit operating costs and increases revenue, creating a positive impact on financial returns.

The next stage up from this is to increase the effective use of the fleet, either by running more frequently, or by coupling vehicles to make consist/trains. Platforms may need to be extended. In the earlier example, a 6-min service required a fleet of ten vehicles. If the round-trip time cannot be reduced, a 5-min service can be provided by using two extra vehicles, increasing the fleet to 12 (20% increase). Alternately, if the running time is reduced by about 10%, then a more frequent service may be possible. The most frequent urban light rail service historically was every 60 s, but with other road traffic, the practical limit is now probably about every 2 min. There is still considerable scope to increase frequency, capacity and hence patronage on an existing line.

Reducing the layover time at terminals can be another way to increase frequency. This, however, needs extra crews at the terminal. The incoming crew has a break and its vehicle is immediately taken out by a relief crew that brought in an earlier service. This stepping back of crews requires good rostering and supervision, since a replacement crew arriving late will severely disrupt the service.

5.6.2 New routes

New routes offer the tramway operator both a challenge and an opportunity. If there is only one rail line, adding a second increases the journey opportunities (with a possible interchange) geometrically available. This is network synergy. In building a second and then a third line, there will be lessons learned from the first. These may cover station design, track layouts and other operational aspects which can be incorporated in the new line. The object will be either to save costs or improve passenger comfort or service. Better still, both can be addressed.

The three main financial aspects of a tramway extension (capital costs, operating costs and revenue) will have been confirmed by the recalibrated financial models. This will give even better levels of prediction, so that the light rail operator can promote a second line with considerably more confidence than the first. This will also mean that funders should be willing to offer better terms, since they too will be more confident of an attractive return on their investment.

Further the public and highway authorities will better understand what a light rail system requires and can provide in terms of additional passenger capacity and the reduction of car trips, traffic and congestion. Having seen the performance of the first line, especially if significant car trips have been attracted and traffic reduced, the highway authority will usually allow more priority over other vehicles. The operator can also plan for more convenient locations for stops.

Indeed, there may even be private developers willing to fund a new station integrated with his development, as occurred at Ludwigshafen (Figure 3.12). Here a department store provided a new rail stop within the store, making it very easy for passengers to use that shop. Other possible areas for similar co-operation are railway stations and airports, where the light rail can feed or distribute traffic to and from those hubs, allowing passenger volume growth without having to add to automobile traffic or more parking space.

Usually the biggest operational question is whether the rail vehicles on the new line can use the existing depot or if a new depot (site) is needed. If this is the case, then the new depot might only provide stabling and vehicle valeting facilities, with trams going to the original depot for maintenance or repair, creating economies of scale and, therefore, unit cost savings. On the other hand, this would give the operator the chance to build a state-of-the-art maintenance facility and demote the original depot for purely overnight stabling.

5.6.3 New service pattern

A second (or third) line will provide the chance to introduce a new service pattern, building on the experience of the first line. There are basically two route strategies. Many to many at infrequent intervals, or few to few at more frequent intervals. The market research on passenger behaviour shows high sensitivity to (perceived) waiting time and hence service frequency.

A convenient rule of thumb is that the service frequency, and thus passenger waiting time, should be less than the in-vehicle time if it is to be an attractive service (Cudahy, 1995). A light rail line with an outer terminal to city centre journey time of 15 min will get few passengers if the service frequency is every 20 min. Ideally, the service frequency should be about 30% of the in-vehicle journey time. In this example, patronage will be maximised at a frequency of every 5 min or better.

When a new line is added, or a line extended, a careful reconstruction of the timetable is needed. This should not disturb the existing travel pattern. Exploring the opportunities for interchange,

Figure 5.33 Light rail/trolley coach interchange station Zurich

Source: City Transport Information

including new through routes, is needed as part of the initial planning. At the very basic, changes to signing and passenger information will be necessary.

If the operator is planning high-frequency services, this will dictate a few to few route pattern. Maximising the journey opportunities by such frequent services means having a dedicated interchange station(s), like the example in Zurich (Figure 5.33). Here, connecting timings allow passengers to alight from one vehicle and almost immediately board the connecting service. At some interchange stations, like the example in Zurich, passengers can get a drink, something to eat or buy a periodical.

Even if this level of timetable reliability is not possible, with a high frequency – for example, every 5 min – then the maximum interchange time will be under 5 min and the average time about 2.5 min. Drivers will also be trained to ensure, where possible, connections are made. There is nothing more frustrating for interchange passengers, seeing the connecting vehicle just pulling away, when a few seconds wait would have sufficed. Staff training will include waiting for interchange passengers. Therefore, the design and location of new interchange stations is critical in the building of a new line. The interchange might be where the two lines intersect, or at a local district centre, sometimes near the central area, or if the new line is a branch off the existing line, at the junction station.

5.6.4 Intensifying rolling stock performance

Most tramway systems aim to offer a consistent and high frequency service throughout weekdays, although daily starting and finishing times may vary. On many systems, Saturday is now the busiest travel day and may need additional capacity including multiple unit operation. The use of twin car sets increases capacity without increasing staff costs.

Once a new system is established and drivers confident in operating the service, three ways exist to speed up the timetable. This is important to minimise costs through productivity improvement, getting more journeys per day from each vehicle. When light rail vehicles are stopped, there are costs and little revenue is generated. Reducing times spent at rail stops to allow passengers to get off or board increases the system efficiency. Drivers can use the passenger address system to encourage passengers to use all available doors and therefore reduce stop dwell time. Reducing or eliminating ticket purchases from the driver using machines on or off board and an honesty system combined with random ticket checking can significantly reduce stop time for little capital investment and with an acceptable level of revenue.

Traffic signals and traffic congestion also delay light rail services. In co-operation with the local highway authority, measures to give light rail vehicles priority over other traffic will reduce running times and therefore make vehicles more productive. Within the parameters of existing speed limits, it can also be possible to have faster running times by faster rates of acceleration and braking (Figure 2.30). These, together with the above methods, can also reduce running times. The maximum rates of acceleration/braking will depend on the traction equipment and the passengers. The highest acceleration that standing passengers can comfortably accommodate is 1.8 m/s^2 (6 ft/s^2). Typical European LRVs are set for 1.2 m/s^2 (4 ft/s^2).

Finally, terminal times can also be reduced by stepping crews back, leaving enough time for crews to swap vehicles. This means that the light rail vehicle can return immediately, while the crew has a break. This requires at least one more crew than the number of vehicles on the route, two if the same is done at the other terminus.

As an example, if the existing round-trip time is 60 min with a tram every 6 min, reducing the round trip time to 50 min means that a service every 5 min can be offered with the existing fleet. Computer modelling can confirm the practicality of new running times within the normal variability and probability of more passengers, or traffic delays.

Indeed in the initial phase, where the patronage generated by a 6-min service is now carried by a 5-min service, station dwell times can be reduced further since fewer passengers, about 17% less, will board or alight, on average, each time at each stop. This might go down to the minimum needed to open and close doors, further speeding up the service. On the other hand, a more frequent and slightly faster service will generate more patronage, which might slightly increase the stop dwell time back to the original but with a larger patronage.

Patronage at different stops will vary. Some will be busier than others. The busier stops can be designated as 'compulsory' and the others request. This will reduce the total running time, without the need to stop where there are no passengers wishing to alight or board. This variability can be modelled as part of the timetable construction exercise. This does however require trams to have an internal system for alighting passengers to advise drivers.

There are other ways to improve rolling stock performance. Normally, the morning and evening peaks require the full fleet to be available. Rolling stock has to be maintained. If minor maintenance can be scheduled for the inter peak period, or overnight, then the full fleet can be used during the peak and the 'spares' taken out of service during the off peak for routine maintenance.

5.7. Engineering for freight on light rail

The carriage of freight on a light rail system needs careful consideration. Freight operation must not detract from the safety or reliability of the passenger service. Special vehicles for freight may be needed and ideally should be loaded and unloaded off the main passenger lines, as happens in Dresden with the Volkswagen trams running between factories.

For light rail systems which share existing railroad tracks used by freight trains – for example, San Diego – the joint track use will normally be time separated, since freight trains run more slowly and take longer to pass. Freight trains running overnight will benefit as there will be no light rail service. The engineering adaptions required will be to ensure that the light rail infrastructure does not foul freight train movements or the opposite. This particularly relates to stop platforms, as light rail cars will normally be narrower than rail wagons. One way to achieve this is to have light rail stops on a side loop off the main line. This, however, will require a consideration of different wheel profiles and track crossings, where an LRV wheel may find rail crossing gaps causing problems.

Where a light rail system is going to be adapted to carry freight, there may be more obstacles, including street furniture that will have to be cleared. Contemporary light rail vehicles are wider than maritime (ISO) containers, so these can be carried on adapted light rail vehicles, provided there is sufficient clearance on curves. If a 12 m (40 ft) container can sit on a modified LRV (Figure 5.34) which clears the curves, then there will be few problems. Special trams can also move the biggest ISO containers (16 m (52 ft) by 3.4 m (10.5 ft)) (Figure 5.35). Presently in many places these maximum sized ISO containers can only be moved on road vehicles.

Figure 5.34 Cargo tram hauling two containers on old trams, Zurich

Source: Neil Pulling

Figure 5.35 Freight tram for maximum size ISO containers

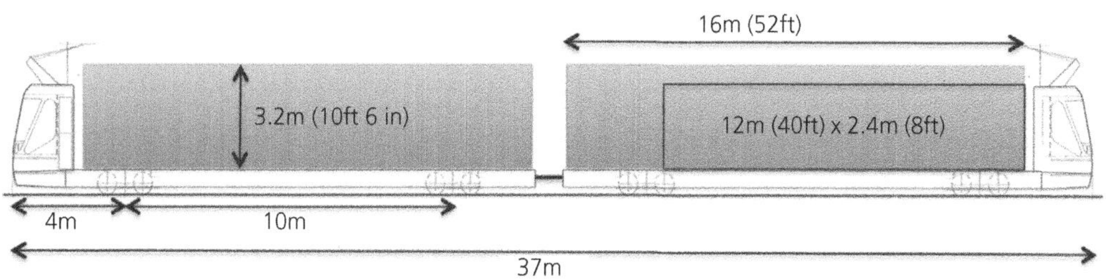

Source: The author

For smaller consignments, the wheeled secure cage/pallet is probably the most versatile unit in widespread use. This can be wheeled on/off light rail vehicles, with level platform boarding. Given the use of web-based freight delivery schedules, the consignment can be met at the stop by arrangement, then wheeled to a nearby shop, store or office as appropriate and empty units returned to the consolidation depot for refilling.

The engineering issues for the stops are that any ramps should be gentle, as a wheeled pallet of soft drinks can weigh a tonne and that access/exit routes are wide enough to allow passengers to walk to or from the platform without being obstructed by wheeled cages being moved. The literature on warehouse and depot design includes dimensions, radii and gradients to guide the designers of new light rail stop platforms.

5.8. Conclusions

This chapter has set out most of the things that are important in transforming the designs of a (new) tramway into practice. These considerations include liaising with approval bodies, properly briefing the contractors that will undertake physical works, or provide necessary equipment. As importantly will be full liaison with the public, since some works may cause temporary disruption, and getting feedback on how things might be undertaken better or in minimising inconvenience. Finally, having good links with the local media, which will also include the use of the internet to transmit information and news as well as gaining intelligence from the local community, is recommended. This is crucial since although the local community will benefit from the new tram system, it may also be inconvenienced, particularly during construction.

emerald PUBLISHING ice Publishing

LJS Lesley
ISBN 978-1-83708-399-2
https://doi.org/10.1108/978-1-83708-398-520261007

Chapter 6
Economics

This chapter discusses both the finances and economics of new tramway or light rail systems, and means to make systems viable. The difference between the two relates, in the first instance, only to those aspects where actual money is transacted. In the second instance, it also includes those social, environmental and related impacts for which there may be no money transaction.

6.1. Introduction

Economics is the science of determining the relationship between the use of money and the utility of resources. Economics includes not just the direct capital and other financial costs but also indirect and non-financial impacts. The discipline of economics has been greatly expanded over the last 100 years but in terms of assessing the rationale of investing in a new light rail scheme, a lot depends on subjective judgement or opinion. This brings such projects firmly into the realm of politics, especially when the investment depends on public funds, since there will be competition with other public sectors – for example, defence, health, law and order, education and so on. Even within the transport sector, a light rail scheme will compete, for example, with funding needed for the maintenance of roads.

What is clear is that assessing a new light rail scheme based on financial costs alone is less open to dispute, except perhaps over the patronage that might be attracted and, hence, the revenue generated. Normally, such assessments are on the basis of 'discounted cash flow' (DCF), whereby income today has a higher value than future revenues and expenditure in the future has more benefit than that needed now. This is an important input for benefit-cost analysis (BCA). The mechanism for these financial adjustments is by way of the discount rate, usually the commercial interest rate charged by banks. The main question is how long should the discounting period last? For many commercial organisations, a maximum of ten years period is often used, on the basis that such investments may be worn out or obsolete. For a light rail scheme with at least a 30 year initial life, ten years may be too short. This would affect the overall test, namely that the discounted income should exceed the discounted costs by a margin. The length of the discount period is another subjective decision.

A fuller discussion on project evaluation can be found in Section 6.7 below.

6.2. Capital costs

The capital cost of a light rail system includes the preparatory works, all the physical infrastructure (including track and power system), the vehicles, staff training and any working capital needed to begin operation until the revenue exceeds the operating costs. These costs have been criticised in the US, UK and elsewhere. The UK National Audit Office (2004) wondered why projects in England often overran both their initial capital cost estimate and the time taken for construction. In real terms, these cost overruns were sometimes 100% higher than equivalent schemes in other European countries. The promoters of publicly funded projects, often a transport authority, can sometimes be seduced by the availability of 'free' capital grants from government. This can lead to grandiose plans, allowing for all possible future options and to 'gold plate' the infrastructure or vehicles. Contractors bidding for publicly funded light rail works also benefit from the budget normally being in the public domain. It can be no surprise that bids often are up to or above the available funding.

Some critics compare the cost of new light rail vehicles with the equivalent capacity of buses. The difference used to be that LRVs were five times the cost of buses (Lesley, 2005). Today the difference is nearer ten times. In practice, trams have a book life of 30 years and, historically, many have been in service longer. In contrast buses only have a 12–15 year book life and often in practice less. The latest generation of trams can easily carry 200 passengers and have a working life of at least 30 years, whereas even articulated buses with mostly standing can only carry 120 passengers.

Intrinsically, buses and trams are equally complex vehicles but of course there are many more bus builders than rail car builders, so market competition is less strong for the supply of trams. Using a supply-side analysis forgets the impact of perception by those who should provide the demand by being passengers and, hence, income.

Open bidding might be a sensible strategy if public money was unlimited. For the Government with limited (capital) funds, a new light rail system will be weighed in the political calculus against the opportunity cost of delivering other elector benefits, including tax cuts. In the UK, a new tramway costs the same as some district hospitals or many new secondary schools. It is, therefore, not surprising that, in the UK, no new schemes have been funded by the Government since 1999 (Nottingham). Ambitious projects in Leeds, Liverpool and Portsmouth were rejected in 2005 and in London in 2008. Subsequently a revised scheme in Leeds was rejected in 2026. Extensions of the Manchester Metrolink and Nottingham have been funded locally, and schemes in London cannot expect any public funding except for asset renewals.

The discussion on value for money in light rail goes back nearly 40 years (Lesley, 1987). Two ways to reduce the capital cost of light rail systems were then discussed. The first by building only what is actually needed to get a system up and running, and the second by financial engineering, especially by having a short construction period for passenger revenue to be generated early, so reducing any outstanding debt more quickly. In net present value terms, the long term counts for little when a discount rate of 7% is applied.

The results are highly sensitive even to small changes in the discount rate used for benefit-cost analysis. A light rail vehicle costs about ten times as much as a bus. To use such vehicles with money costing 15% is a nonsense, and was even more difficult to justify between 2009 and 2022 when UK bank rates were about 1% per year; since then they have been about 5% per year, reducing only recently to about 4%. Governments often also distribute capital grants over a number of years, so that schemes that could be built in a year often take much longer, ensuring that the pain of construction disruption is maximised.

To achieve affordable capital costs needs clear project objectives as discussed in Chapter 3. This is another area of criticism by the UK National Audit Office (NAO) 2004 report. Public authorities are tempted to think that new light rail systems can magically regenerate an economically deprived area or reduce social exclusion, confusing cause and effect. Today the UK political buzz word is 'connectivity'.

New transport systems should target diverting existing car traffic flows, especially as cars cause congestion, and traffic pollution causes injuries and deaths. Getting enough passengers to pay for the operating costs is another challenge for system promoters, especially if fares are set for political reasons rather than with reference to the local travel market – for example, taxi fares.

The most direct light rail route is likely to have the lowest capital cost and, by providing the shortest ride time, be most attractive and therefore divert the maximum number of passengers from cars. An example of this is comparing the first Metrolink line in Manchester (between Altrincham and Bury), with the second (between Cornbrook and Eccles). The first line is direct, fast and generated an operating profit of some £5 million per year. The second meanders around what was the 'brownfield' Salford Quays and is slower than buses running on street. Not surprisingly, this line reported an operating loss of about £10 million per year.

After a direct route, avoiding the need for tunnels and bridges should be high on the promoter's agenda. Not only are these expensive and slow to build, but they impose operational constraints and are costly to maintain. Often the structures are justified in avoiding road traffic problems. An example of this was in Shandwick Place in Edinburgh, when city planners demanded that a new tramway should be in a tunnel to avoid disrupting car traffic. In most city streets where a light rail route will pass, every LRV will stop at nearly every stop to allow passengers to board and alight. In the same city street, over 99% of car traffic will travel non-stop. The logic, therefore, is that if a subway is needed to separate light rail from other traffic, the other traffic should be diverted into a subway and the light rail operate at grade on street, maximising the convenience for passengers and the attraction of trips from cars.

This, however, is where an enterprising promoter will face head on road traffic problems, by designing traffic management measures that give priority to LRVs through time-sharing traffic signal plans and space sharing with exclusive tram lanes. This is a more economic option than subways and allows more stops to be afforded and located conveniently for passengers. It may also require simulation modelling to persuade city politicians and highway engineers that light rail can operate without adding to congestion, or even reduce congestion by attracting of car trips. This is much less costly than building tunnels and less disruptive to traffic and the local economy during construction. The light rail service in San Diego demonstrates this in several ways. Figure 6.1 shows the peak patronage of the original bus service, then that of the replacement light rail line, immediately after opening and several years later. The extra passengers have been attracted from car commuters.

The third area in which to achieve more economic capital cost is in the stops. Not being in tunnel or on viaducts, they will immediately be less costly to build and give flexibility in siting to get closest to the main traffic generators and attractors. Stops should be built to handle only the vehicles that will operate when the system opens. At some stage, if coupled operation with two car sets or longer articulated vehicles is planned, then all stops should be designed so that platforms can be extended. Extra accesses will also have to be available for the larger passenger flows generated at the busiest stops.

It may also make sense to build one or two of the busiest stops with longer platforms from the start, so that two LRVs can occupy the platform at the same time and provide a model from which future platform extensions can be based. Patronage forecasts will also predict the likely daily flows at each stop. This will help determine the most economical way to provide for these flows, especially the peak period flows – for example, determining the width of platforms, while maintaining a safe system and corporate style for all stops.

Figure 6.1 LRT ridership in San Diego

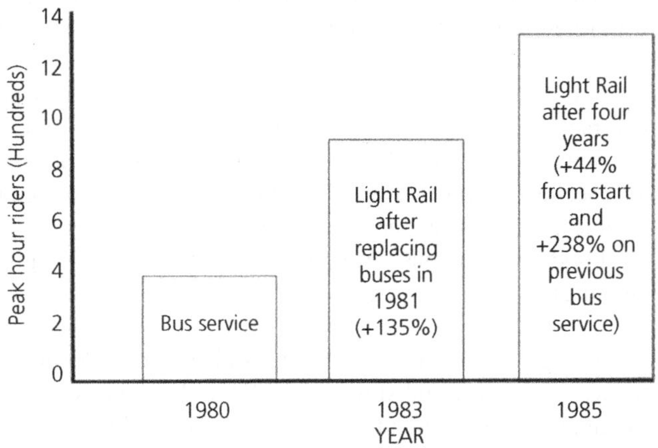

Source: San Diego Transit

Finally, designing and building the infrastructure as simply as possible will reduce costs and give fewest future maintenance problems. The Merseytram project, aborted by the British government in 2006, planned four delta double track junctions in the city centre where, following the above analysis, none should have been needed. This would have saved probably 10% of the track capital cost and considerable construction disruption.

The depot is another major capital investment. Promoters should avoid the temptation to build for any ultimate system size. Providing only for the initial fleet in terms of stabling, valeting and maintenance is the most economic option, provided land can be reserved for any future extensions. Not saddling the initial system with the capital costs appropriate for one much larger will also help to make it affordable.

As a poor example, the planned Merseytram depot at Gillmoss for 18 LRVs, would have been twice as big as the Merseyrail Kirkdale depot, stabling some 50 Class 777 train sets. The capital cost of the depot can be further reduced, if the vehicle supplier has a maintenance contract. The system promoter can then trade off the savings in capital expenditure on maintenance equipment, against the extra revenue cost of contracting out the maintenance.

The above discussion is directed at systems that are publicly funded. Raising the capital for privately funded systems is very different. Historically, virtually all tramways and railroads were privately funded. Most were built by contractors who were not paid wholly in cash by the promoting company but in a mixture of cash and shares. On completion of the project, the company was then floated on a stock exchange and the contractor sold his shares on the stock market, an early form of the 'dot-com' option and generated more money than would have been received from a straight cash contract. This method of financing projects lasted about 80 years.

Today, contractors usually want to be paid in cash, so the promoting company has to raise all the capital needed upfront. Normally, however, a contractor may be one of a consortium that is promoting the project and will be required to put in some capital, either in terms of funding, or staff and other resources. The contractor will also be expected to charge an 'at cost' price, in return for profit sharing.

Usually, some of the capital will come from investment companies but the bulk is often raised from merchant banks. These, however, expect that the promotors invest at least 10% of the initial capital. Even merchant banks will divide the capital between a syndicate of other banks – for example, in constructing the Channel Tunnel. For the more modest capital required for a new light rail system, the merchant bank may also advise on a share issue.

Finally, the promoting company will have raised the capital with a combination of loans (debt) and shares (equity) to construct the new light rail system. Ideally, the more shares there are, the easier it is to achieve profitability. Interest on loans will have first call on any operating revenue. Shareholders will only get dividends when the company is profitable. Once the system is open and revenue comes in, the dividends can be increased by refinancing the debt at lower interest rates, due to the reduced risks from an open and operating system, and the asset created providing security. Further shares can then be issued to pay off some or all of the debt.

Banks will require the interest to be paid on loans, even if the company is not yet profitable. This leads to the company having to borrow more money or starting with a larger loan. The Channel Tunnel company paid as much to the banks in interest to fund the project as they did to the contractors who built it.

Having most of the capital raised from shares does mean, however, that the share price must be kept up if listed on the stock exchange, so that shareholders can realise their investments by selling to other investors. Again the Channel Tunnel project provides an example, albeit extreme, of the importance of share price. Investors paid £5 per share, half way through construction and four years before opening, the shares reached £18 each. Four years after opening when it was clear that traffic projections were over optimistic, the shares sank to £0.50 each.

At this point it became clear that no more money could be raised from shareholders and that the bank debts could not be serviced from revenue. A number of refinancing deals were organised, where the banks exchanged their loans for shares, diluting the control of the original shareholders, who today in total own less than 10% of the company. The Channel Tunnel is an extreme example of a privately funded rail project.

The capital needed for a light rail system is much smaller in comparison and the risks less. It should be fundable from investors in the local area, who will benefit either as users or in trade terms. Local investors can be offered free travel as an extra reward on top of dividends.

The top priority of such private projects is the maintenance of confidence. This can be achieved by the use of reputable contractors and as importantly by competent management with good public relation skills. Meeting published deadlines is critical to keeping up confidence. Once one line has been completed as expected and revenue service starts, then the promoters will be able raise further private capital more easily.

Charles Yerkes made a fortune from the North Chicago City Railway, creating a large transportation empire. He then built up the Underground Group in London, in the early twentieth century.

This issued new shares for each new tube line built. By this means, the bulk of the underground system in London was constructed. Similarly, the first generation of streetcar and tram systems were privately funded, until the Wall Street crash made raising new funds from shares difficult or impossible. In England, the British Electric Traction Ltd (BET) built nearly half the country's tramways. BET still exists but is a general service company, having had its transport assets nationalised in 1968.

Some new light rail systems do not own their own fleets, leasing them from a bank or other financial institution. This can be more economical than buying the fleet outright, and certainly reduces the upfront capital investment, but depends on the local taxation rules and credits. In those countries where there are tax advantages for large financial institutions to offset the lease income against the capital cost, then leasing can be an attractive option.

A lease and maintain deal can be part of the financing of a new system. A lease deal can be paid for on the basis of the availability of vehicles. The operator can fine tune the revenue generated to meet fully the cost of rolling stock. The disadvantage of a leasing arrangement is that the leaser will always wish to have a fall back in case of default, by being able to use the vehicles on another system. This will restrict the operator in the amount of local customisation of the trams. On the other hand, a vehicle lease deal that includes mid-life refurbishment, with a rental review, can be an equally attractive way to operate and modernise the fleet without capital investment, albeit with payments out of revenue.

The lease premiums will appear as an operating cost and can be set against any company tax payable at the end of the accounting year. This will also quicken the date when the system is genuinely in surplus and shows the importance that financial engineering can make in minimising the capital debt (servicing) and therefore the time before shareholders can be rewarded for their investment risk.

6.3. Operating costs

Operating costs include all consumables and non-capital items needed to operate the service. This includes power consumption, staff wages and salaries, rents, leases, insurance, maintenance, spare parts, other materials and cleaning. Operating costs can be pre-planned in advance of starting operations. Budgets can be further refined during the commissioning period. Normally the biggest cost is staff wages and salaries.

Promotors must resist the temptation to make new light rail operations overly complex. This is one of the reasons that new systems have been uneconomic and heavily criticised by the UK NAO (2004). In 1999, Joan Dunbrack of the Boston Transit Authority (MBTA) was visiting the Supertram system in Sheffield. On being shown the control room, the three members of staff on duty, controlling the vehicle dispatching, the power distribution, the trackage and so on, explained how difficult it was to control a peak fleet of 25 trams. Mrs. Dunbrack listened politely then said that she did the same job as the three controllers, single handed in Boston. She added that the morning peak fleet was 250 trams, which she managed. Keeping the operations simple, not only reduces the operating costs but also the likelihood of things going (badly) wrong (Cuffe, 1989). This should also reduce the capital costs.

The staff of drivers and, if they are used, conductors, will be the largest staff costs. In most systems, the morning and evening weekday peaks will need the largest number staff available. Some operators use a combination of full time staff to cover the base service operated throughout the day and part time staff to cover the morning or evening peaks. In some systems, a hybrid split shift can be used, whereby staff will work both the morning and evening peak, and have most of the morning and afternoon off. There will be a rota to share different shift patterns. Agreements with unions also has an impact on staff costs.

The hourly rate of pay, unless there is a salary contract, will reflect local labour market conditions. The staff needed to operate expensive machines in public environments, where safety is critical, will probably require a pay rate above the local norm for other service industries. This should attract higher calibre staff, unless there is a pool of unemployment when pay rates could be lower. Having a choice of staff will allow the operating manager to attract, recruit and retain a motivated work force. Rail systems find that staff are more stable and loyal, and, on average, stay much longer than bus drivers do. To get the best out of staff, the training regime needs to be more than just the mechanics of a tram service. This can include public relations and providing passenger information.

All staff, together with pay rates, will determine the total staffing costs. Many light rail system in the developed world need about 3 drivers/operating staff on the payroll for every vehicle in the operating fleet. This covers an 18-h or longer working day, holidays, sickness and continual training requirements. Good management and the use of non-pay incentives, as well as pay, can have a large impact on improving and maintaining morale. Included in this can be seniority in the choice of shifts, to encourage staff to stay, reducing turnover and, therefore, recruitment and training costs.

The depot and workshop will be the next largest staff cost centre. For an efficient operation, there should be enough work to keep all the staff fully occupied. For a newly opened system, there will rarely be significant maintenance and, unless drivers are careless, little accident damage to repair. One option adopted by some systems is the use of multi-skilled staff, so that when there are no workshop tasks, staff can be used for other duties, including infrastructure maintenance and operating the system. Indeed there is merit in having all staff, including the CEO, working an operating shift driving a light rail vehicle, at least once a month. In many systems, the depot staff will also be the emergency team to turn out to an accident and assist the emergency services – for example, re-railing or rescuing a crashed or over-turned vehicle.

This theoretical analysis may fail to succeed because of the perception that overall costs are higher than, for example, a bus service. Real data can show a different picture. Analysis of data from US cities with a new light rail system has compared the operating and maintenance costs with bus services in the same places (Table 6.1). In most cases LRT systems have lower operating and maintenance costs per passenger mile than bus; the worst at 90% (Baltimore) and the best 30% (St. Louis), with a weighted average of 64%.

Similarly, data from Nantes, in France, shows that trams generate passenger revenue of 113% of the operating costs, whereas local buses only 53%. Trams in Nantes therefore cross-subsidise bus operations. These figures should give a light rail promoter confidence that, by adopting the best or nearly the best practice, operating costs can be (much) lower than the equivalent bus costs and passenger revenue higher.

Similar data was presented to the UK Parliament by Transport for London, comparing bus and tram operations.

Table 6.1 Bus and LRT operating and maintenance costs in 2001

City	Operating and Maintenance ($million)		Passenger Miles (million)		Cost ($) per Passenger Mile	
	Bus	LRT	Bus	LRT	Bus	LRT
San Diego	58.2	26.5	153.5	152.7	0.38	0.17
St. Louis	98.8	19.2	150.4	95.9	0.66	0.20
L.A.	633.4	51.4	1,332.2	170.5	0.48	0.30
Portland	120.9	22.3	237.0	63.3	0.51	0.35
Sacramento	41.8	14.8	79.5	39.4	0.53	0.38
Dallas	161.0	27.7	218.8	58.9	0.74	0.47
Baltimore	150.2	23.0	285.9	48.0	0.53	0.48
Denver	142.1	8.0	269.8	13.1	0.53	0.61
San Jose	145.9	26.2	202.4	33.0	0.72	0.79
			Weighted	**Average**	**0.53**	**0.34**

Data: www.lightrailnow.org/facts/fa_lrt02.htm

6.4. Patronage, fares and revenue

This is the area of assessment that is most open to alternatives and interpretation. Over-optimistic forecasts are a major criticism of many recently opened light rail systems. From 'where will the passengers come?', a transit executive once asked. The population living in the light rail hinterland is going to generate most of the patronage initially, on which the revenue forecast must be based. These residents already make trips and will be attracted from other modes of transport, mainly private car, as this dominates trip making in most developed urban areas (Figure 6.3). To this can be added, as appropriately, park and ride from outside the catchment, the diversion of trips from other destinations to use the convenience of the light rail system, the generation of trips which previously were not made within the area, and interchange from other public transport modes.

The number of trips made per week per person has stayed remarkably constant over the last 50 years (Figure 6.2), at about 18 in the UK. There is no evidence from elsewhere that trip making is significantly different, or has grown over the past few decades. The reason for this is that going to work (or school) is the dominant journey purpose for many people and there are only so many work days in a week.

What has changed has been the mode of travel and trip length (Figures 6.2 and 6.4), with the work journey more than doubling in length on average and with a major switch from public transport to private cars, so that in 2005 almost 70% of all work journeys in the UK were made by car and only inner London has a significant use of public transport to get to work. Indeed, Central London is the only place in the UK where public transport trips out-number those in private cars.

Figure 6.2 UK passenger behaviour 1972–2024

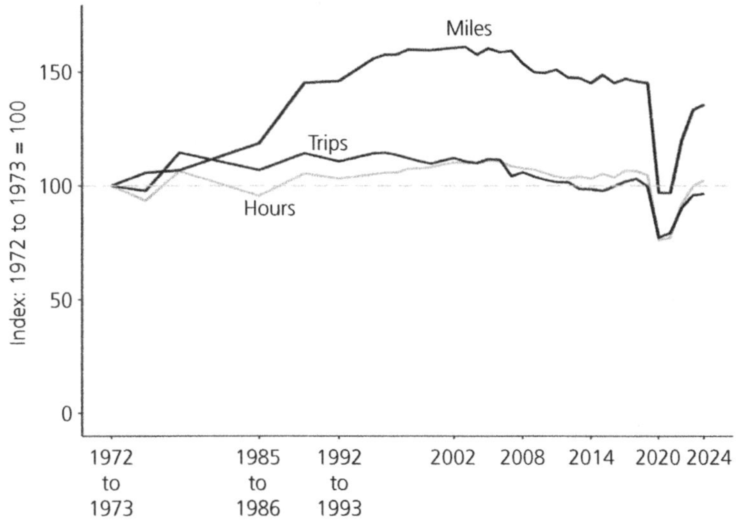

Source: UK Department for Transport – National Travel Survey

Figure 6.3 US travel trends 1960–2018

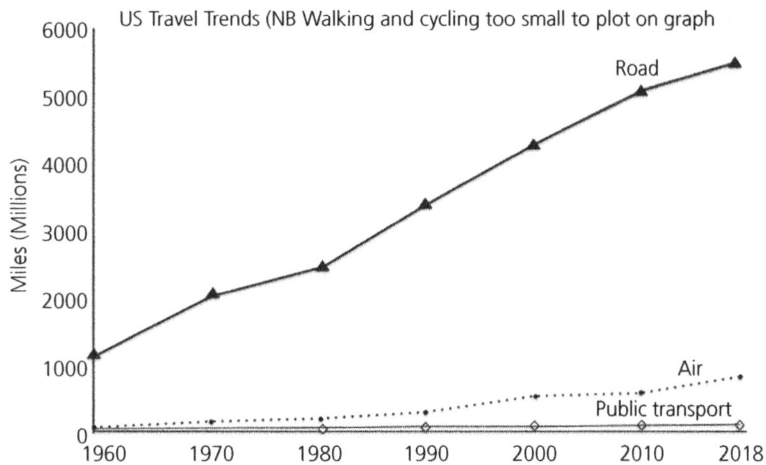

Source: US Department of Transportation

New York accounts for over half of all US rail transit trips and most of those are commuters into Manhattan. In the rest of the US, metropolitan trips are dominated by cars – for example, in Houston over 90% of all trips are by car and less than 5% are by transit. The trip length of US commuters increased from 2001 but has stabilised (Table 6.2) at about 19 km. In comparison commuters

Figure 6.4 UK travel mode share 2018

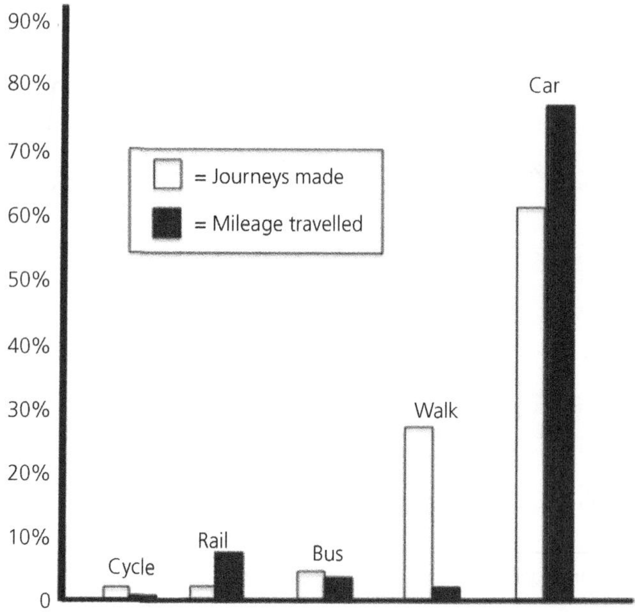

Source: UK Department for Transport

Table 6.2 US commuter trip lengths

Year	Trip Length miles (km)
1983	9 (14.5)
1990	11 (17.7)
1995	12 (19.3)
2001	12 (19.3)
2009	11.8 (18.9)
2015	12 (19.3)
2020	12 (19.3)
2025	12 (19.3)

Source: Federal Highways Administration

in the City of Boulder (Table 6.3) have an average commuter trip length nearly 50% longer than the rest of the US.

Commuters in Europe have shorter trip lengths – in the Netherlands (Table 6.4), almost 40% shorter. In the UK the figure for 2024 was 16 km (9.8 mile), 20% shorter than the USA. Added to this, European residents make more trips by public transport, as in Freiburg (Figure 6.5).

Table 6.3 City of Boulder commuter trip length (Boulder, 2022)

Year	Trip length miles (km)
1990	5.3 (8.5)
1994	6.2 (10.0)
2022	18.0 (29.0)

Table 6.4 Netherlands commuter trip length

Year	Trip Length miles (km)
1985	5.4 (8.7)
2000	6.6 (10.6)
2022	7.5 (12.0)

Figure 6.5 Modal share in Freiburg, Germany, 1990–2020

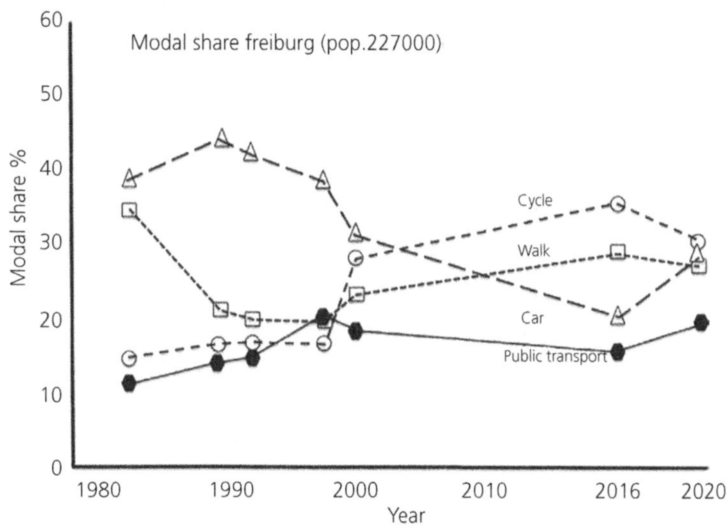

Source: Freiburg City Council

With concern about dependence on fossil fuels (Figure 6.6), environmental pollution and traffic congestion, there should be a slow but growing modal transfer in the US and other developed countries from private car to transit, and particularly light rail. Two reasons for this will be the relative affordability of new light rail lines and their proven ability to attract car trips. The one confusion to

Figure 6.6 Energy use in the USA, EU and UK

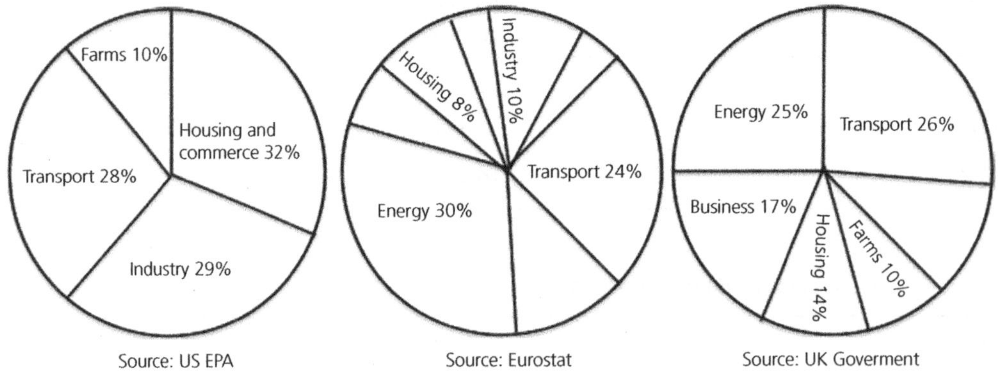

this has been post 2020 (Covid), the number of commuting trips has reduced in some countries as people have taken to working some days at home.

How many of the 18 trips made per week per person in the UK can be attracted to light rail? There are well established economic diversion curves, which will predict the probability of a trip being made by one of several alternatives, including light rail. This is calculated on the basis of generalised costs. These bring together the facets which market research has shown influence mode choice decisions. They include journey time, waiting and walking time, interchange time, comfort, convenience, cost and so on.

These have different impacts on the mode choice decision. They are therefore weighted to bring them all to the same equivalent units of cost or, more usually, time. Time as a constant measure has the advantage that it is monetarily inflation proof and can be used in different countries. It is also more logical, since a shorter generalised time correlates to more trips being made, just like the classic economic price/demand curve.

Part of this assessment is the number of trips that can be attracted to a (new) rail line. The market condition of alternative transport options will also have an impact. These include the cost of car parking, bus frequencies, taxi fares and so on. Indeed any modelling of likely rail patronage will also include sensitivity analyses of how competitive modes of transport might react and therefore impact on tramway patronage.

In two such studies, the impact of competitor bus operators charging zero fares was examined. In both cases the result was a reduction of rail ridership by under 25%. Zero bus fares however cannot be a long term option for a bus operator. In some countries it would be considered to be anti-competitive – for example, illegal in the EU and USA. A bus operator charging zero fares will either go bankrupt, or will then have to raise fares above the original level to replace the lost revenue. In either case, this will benefit the rail operator. This does, however, raise the question of a general city-wide policy of zero public transport fares, as has been implemented in Tallinn and Luxemburg, see below.

The number of rail passengers is also sensitive to the fares charged. For politicians, low transit fares can be a mantra. Little if any evidence exists to show that low fares attracts any passengers from cars. This is because fares are usually less than 25% of the generalised cost. They are also much less than the real financial cost of using a car. Fares, being a small part of generalised costs, have a small impact on mode choice decision making, since 75% is based on service quality. The low fare elasticity, usually between –0.3 and –0.4, confirms this. This means that a 10% fare cut leads to a 4% increase in ridership and so a 6% decrease in revenue. This relationship has been well known in the public transport industry for at least 100 years.

During 1975–1986 in England, South Yorkshire Passenger Transport Authority (SYPTA) froze fares. Between 1980 and 1986 (Hay, 1986), the Merseyside Low fares experiment and the London Fares Fair scheme between 1981 and 1982 followed suit. The low fares regime in London, introduced in 2024, has been abandoned in face of rising operating losses.. The logic behind the SYPTA policy was that with inflation over ten years, the income from fares would fall compared to operating costs, to a point where it would not be worth collecting fares.

Two European cities operate free fares: Luxembourg and Tallinn (Guardian, 2025). In neither city has there been a significant switch of trips from private cars to public transport, although like the earlier experience in Sheffield, people on low incomes make more trips by public transport. This is a large charge to the cities concerned, covering the operating costs and also the capital costs of replacing assets when worn.

There was little statistically significant data to show that low fares attract any trips from cars to bus but many dependent bus users did make more bus trips. In the UK, the most prolific taxi users are in the lowest income quartile of households, which have the lowest level of car ownership. High fares are not a disincentive to taxis use, since cost is balanced against the occasional use and quality of service provided.

Therefore, a (new) rail system will wish to position itself between the fare levels of bus and taxi. For a commercial system, this will be market driver. For a public system, if a political decision is to have low fares, then there will have to be subsidy or other compensation to make up for the lost revenue.

The principle for this has already been established in the EU. In the UK, where pensioners have free bus travel, the bus operators receive compensation on a 'no better/no worse' basis. The operators get the difference between the fares they would have collected from (fewer) pensioners paying the full fare, compared to the actual number of pensioners who travel free. Some operators complain that this is not enough. If concessionary travel makes up for a large part of bus patronage (33% in UK) (UK DfT, 2022), then more buses are needed to provide extra capacity in the afternoon peak. This increases operating costs.

Table 6.5 shows the typical sensitivity to different fare levels on a light rail system. The maximum patronage is with free travel. Increasing fares reduces patronage but by a smaller proportion (elasticity of demand), so revenue increases. The opposite happens when fares are reduced, patronage increases but revenue falls.

The number of passengers and the average fare charged, together produces the total revenue. Again in one commercial project, the sensitivity of patronage and revenue to fare level was

Table 6.5 Revenue sensitivity to rail fare in a new light rail system

Rail Fare	Patronage (million per year)	Revenue (£ million per year)
zero	6.0	0
£1	5.0	5.0
£2	4.5	9.0
£3	4.0	12.0

Source: The author

calculated (Table 6.5). The largest patronage (6 million per year) was achieved with a zero rail fare. Clearly this is not a viable option. The operator should use market research to determine the optimum fare level to generate enough revenue both to cover operating costs and also the depreciation and renewal of assets. By comparison, the equivalent average bus fare in the same town was £1.20 and taxi was £5.50. Today, like low-cost airlines, and with the use of contactless fare collection, fares can be changed during the week, or even during the day, to reflect traffic and travel conditions.

In practice, a commercial light rail operator might offer a variety of fares, perhaps copying low-cost airlines. Maximising revenue yield from low fares for advance bookers and high fares for those who leave it to the last minute or travel at peak times. Another cost model might be that used by cell/mobile phone companies. Indeed cell/mobile phones are an efficient method to collect fares, since they are already being used for such diverse things as paying for parking and retail purchases.

No discussion on revenue collection can be complete without considering passengers travelling fraudulently, without paying. Few if any transit undertakings have ever managed 100% payment compliance. In European countries, many studies have determined the unpaid portion of revenue. Typical figures are between 2% and 5%. Supermarket 'shrinkage' from shop lifting can be higher than 10% of turnover. The light rail operator will have to determine, provided this is a static figure, whether it possible to collect the final 2% of revenue and, if so, if it is worthwhile.

One option is to fit ticket-reading entry and exit gates at stops. This can only be practical on a closed system. There have been experiments in having entry gates to the rail vehicle. In either case, there is a capital investment required and high maintenance costs. These can be reduced by using gates that are normally open and only close if an invalid or no ticket is presented. The rail operator must, however, weigh up the extra revenue collected against making the system less attractive to 'honest' passengers, who can easily use their car instead.

Another revenue protection option is the use of inspectors, either travelling on vehicles or at busy stops to check tickets, as a deterrent against fraudulent travel. Many German rail systems use a sampling rate of one in ten journeys with a 'penalty' fare of ten times the normal for prepaid fare being charged, to those without valid tickets. In a new rail system, an approach of making it easy to pay the right fare and the use of social pressures can be as effective as a revenue control inspectorates, which is another operating cost and can have a negative impact on patronage if perceived to be a 'gestapo' by the public.

Indeed, in 1995 the transport undertaking in Brussels was concerned at a slowly declining patronage and undertook market research to find out why people were travelling less by transit (bus and rail). One of the reasons given was the sinister appearance of the revenue inspectors, nick-named the 'gestapo' by many people. A clever PR exercise included a new uniform, less military in aspect and a city wide poster campaign. This had pictures of all the revenue inspectors, with slogans like 'Uncle John, here to help you get home safely'. After six months, the image of Brussels transport had changed significantly for the better, and after nine months patronage (and revenue) started to grow again.

Similarly, when the author worked in Budapest for nearly a year, revenue protection was by snap inspections by plain clothed staff. During this time only four occurrences of ticketless travel were observed, of which only two were really fraudulent travel.

An operator who forgets that the image of the rail system is as important as the hardware and training is in danger of having no passengers. In 1986, when bus services were deregulated in England, a new bus company (BEELINE) was set up in Manchester (Figure 6.7). Roadside posters along the planned route promoted the new service using mini buses. The established bus company, Greater Manchester Transport, reacted by taking (expensive) adverts on local television. A follow up public recognition survey was undertaken. All respondents remembered the bus adverts on television. Most thought they were on behalf of the BEELINE company. Not only is image important, it has to be developed and defended. The choice of the wrong media can be an expensive mistake. Recently the Manchester Transport Authority has adopted 'Bee Network' (Figure 6.7).

Figure 6.7 BEELINE BUZZ Co. 1986 and 'Bee Network' bus launched in 2023

Source: BEELINE BUZZ and TfGM

6.5. Financial viability

At the very basic level, financial viability means that passenger revenue should exceed operating costs. For all light rail schemes, the operating surplus needs to fund repairs and refurbishment, and replacement of assets when life expired. A public-sector light rail system may have its fares set below the market rate, for political reasons, but will then need a subsidy in compensation. Often this compensation is rarely adequate and operators have to make do and mend, or run down the service (Cudahy, 1995).

An earlier study compared municipal and commercial tram service accounting. Yearsley (1993) showed that the commercial tramways, fully depreciated their assets and paid for replacements from retained revenue. In contrast, the largest municipal operator (London County Council (LCC)) used revenue surpluses to reduce property taxes. When assets needed replacement, this was treated as a grant from council funds and led to political charges of subsidisation. Yearsley showed that had the LCC operated its tramway commercially, on average a dividend of 4% per year could have been paid and replaced assets out of retained revenue.

For a commercial light rail system, an operating surplus will also need to service the capital cost of building and equipping the system, whether this is by loan/debt, equity or most probably a combination of the two. The operator, therefore, has to, in the short term, fine tune the fares to enhance their revenue yield. In the medium term, the impact of patronage volumes will effect fleet size, staff numbers and so on, all of which can improve the profitability of operations. Even in public sector schemes, fares may be raised in the peak period to squeeze marginal traffic onto the off peak, which would otherwise require additional peak vehicles. The extra revenue could be used to fund the purchase of more vehicles.

Monitoring costs and income should be a daily, or at most weekly, exercise. In some parts of the retail industry – for example, food – stocking is updated daily, and even in fast moving fashion and clothing, weekly. The light rail operator needs to be able to take corrective action quickly. Should costs begin to rise, or more likely revenue drop, outside the budget, delaying action can make the eventual decisions expensive to introduce. A crisis leads to a loss of morale in the staff and credibility by the public. Therefore, rapid steps taken to avoid a crisis benefits everyone.

Financial 'engineering' at this stage is as important as the physical engineering. Managing the cash flow, from which an ultimate operating surplus is derived is a skilled task, especially as the expenditure is lumpy and the revenue flow is relatively smooth. Investing surpluses for later draw down – for example, for mid-life refurbishments – could generate nearly half the total required from earned interest, assuming continuing relatively low levels of monetary inflation and interest rate stability.

Monitoring patronage will be an ongoing task for the traffic manager. Like low-cost airlines, this requires suitable strategies to cope with any changes of demand. One option is joint exercises with important traffic generators to target people going to entertainment, shopping and so on. Market research can identify what is being done right and what needs to be improved. By definition all organisations can do better. Feedback from operating staff should also be used, since they are in daily contact with customers, who buy the tickets that pay their wages. The flip side is positive customer care, which can have an impact on reputation and patronage, through personal recommendation to non-riding family, friends and neighbours.

6.6. Project funding

6.6.1 Public promoter

For a publicly funded project, there may be several sources of funding. National or state government can provide grants which may be paid in instalments against project progress. The Government can also give a low interest loan; again this will be paid against project progress, but the interest will also have to be serviced and the loan paid back from operating surpluses, or another source of revenue – for example, local taxes.

Central Government can give and underwrite local government credits for loans from commercial banks, or extra national funders like the World Bank, IMF, investment banks and so on. Again, the interest and capital will have to be repaid but with the advantage of special low interest rate, because risk is low associated with publicly promoted projects backed by government.

Some governments have used complex schemes like private finance initiatives (PFI), or public private partnerships (PPP) to raise the capital costs from private lenders. In both of these, the government, on paper, transfers the risks of the project to a private consortium of constructor, equipper and operator. The consortium raises private investment capital to build and operate the system. Any risk is priced into the consortium's bid price, thus inflating it, to the public promoter and government. In return the government gives the consortium a concession for a number of years, long enough for the original investment to be rewarded. Where an operating profit cannot be guaranteed, the government will offer an annual subvention to underwrite the consortium.

The advantage to governments of such schemes is that the capital debt of the project does not appear against the government's total debt, for the purposes of fiscal policy or to satisfy external bodies like the IMF. This is supposed to show that the government is not living above its means. In practice it is a sophisticated mortgage, for which the government is the fall back. Ultimately, the total cost must be repaid, either by the consortium from passenger revenues or by the government through subventions.

This was well illustrated in summer of 2007 when one of London Tube PPP consortia went bankrupt. The project with extra costs had to be taken back by the public sector to ensure that underground services kept running and the needed modernisation continued. Additionally, the consortium will have front loaded the proposal price to provide for contingences like any changes in public policy.

Private consortia will also develop strategies for increasing the profitability of such contracts. One way in which this is achieved is by re-financing the project when the infrastructure has been completed. The logic for this is that the initial investment is one of very high risk, with a high rate of interest from private lenders. Once the project has been completed and revenue services commence, the infrastructure becomes the security against which to borrow at lower rates of interest and the income stream provides further security.

This allows the original investment to be paid back, which is usually put in by one or more members of the consortium. Members of the consortium thus make a trade profit from providing for the building/equipping the system and then a financial profit of getting their capital investment back with a premium. There are also examples where the new consortium has then sold the concession to another investor, also making a further profit. In the end, such consortia can earn over a 100% return on the original capital invested. Set against this, the public authority pays a larger sum than would have been the case had the project been funded directly from the public purse.

6.6.2 Private promoter and funding

Historically, most transport infrastructure has been promoted, funded, built and operated by private companies. Many transport companies were subsequently taken into public ownership for a variety of reasons, including political ideology, national security, consumer and passenger protection and so on.

A private promoter approaching a new tramway as a commercial venture is seeking to create a profitable business and be able to service the initial capital costs. A private operator will also wish to build up the value of the asset, against which further investments can be generated. A private light rail promoter will therefore seek routes which give the maximum revenue for a minimum capital investment. How will the funding for such enterprises be raised?

Historically, most private transport systems were built with share capital (Figure 6.8). Many contractors were paid in shares. Once the system was built, the company was floated on the stock market. The original investors could then sell their shares to achieve returns from their investments. More recently the dot.com investment boom of the early twenty-first century was funded in a similar way. Like rail in the nineteenth century, this was also based on high tech cutting edge new companies, formed to exploit the new opportunities of the internet and 'e' trading.

A private promoter will most likely use loans and shares (debt and equity) to fund the project. Indeed, any financial institution willing to lend for a new light rail scheme will almost certainly expect the promoter to put up at least 10% of the investment costs and take shares. Merchant banks are one source of such finance. There are also extra-national institutions, like the World Bank, IMF and European Investment Bank, that can give loans at below high-risk commercial interest rates. These help promote better 'public' infrastructure and facilities. This has been one of the ways in which the privatised water companies in England have funded the modernisation of water supply systems, after years of under investment while in public ownership.

Whatever way a private promoter funds the building of the system, once it is open and revenue is generated, other investors can assess the profitability of the project. The promoter can then float the company on the stock market to attract further funds to expand and intensify the service. The advantage of shares as a source of investment capital is that there is no permanent debt servicing. Debt servicing reduces the profits, part of which need to be paid to shareholders as dividends.

Figure 6.8 Nineteenth century tramway share certificates

Source: Archive Websites

Figure 6.9 Commercial modernisation in the UK (1931) and US (1934)

Source: London United Tramways Ltd and US Info Service

A small change in revenue, or operating costs, can make a significant change in profitability and hence dividends. Included in the operating costs will be asset replacement to maintain a safe and modern service. Figure 6.9 shows examples of private light rail companies in the 1930s upgrading their rolling stock.

The promise of dividends on shares, when a company is profitable, may be more attractive than interest on a loan. There is a second benefit from a capital gain, if the shares are quoted on the stock market. Historically, company shares have appreciated in monetary value more than other commodities. So even if a company pays little in the way of dividends, shareholders can be more than pleased by the rise in value of their shares on the stock market. Depending on the local tax regime, this might be financially more attractive to investors. Even an event like the 2009 'credit crunch' fall in stock market values is likely to be short lived and investor confidence will return allowing new projects to be funded.

The other major attraction for a private promoter of raising the needed capital by means of shares, is that, unlike a loan, the capital costs do not need to be repaid. The shareholders can recover their original investment by selling their shares to other investors, or indeed having them bought by another company wanting to take over the rail company. For these reasons, therefore, a private rail promoter will always try to maximise the capital funding by way of shares and keep the value of the shares high and growing, even if dividends are not always high.

6.6.3 Financial 'engineering'

A promoter can also make the investment more efficient by changing the construction time and schedule. If interest rates are high this approach can lead to significant cost savings. The tables below illustrated some of these principles for a project with a total cost of £240 million. These show different timescales for completion and different rates of interest. They show how sensitive is the final cost and the benefits of a shorter construction and completion time.

The option in Table 6.6 with interest comes to a total capital debt of £322 million and an annual debt charge of £23 million.

Table 6.6 Ten-year construction period and 7% interest per year

Year	Costs: Infrastructure	Vehicles
1	20	0
2	20	0
3	20	10
4	20	0
5	20	0
6	20	10
7	20	0
8	20	10
9	20	0
10	20	10
TOTAL	**200**	**40**

Table 6.7 Five-year construction period and 7% interest per year

Year	Costs: Infrastructure	Vehicles
1	40	0
2	40	10
3	40	10
4	40	10
5	40	10
TOTAL	**200**	**40**

Table 6.7 shows that with interest the total debt comes to £292 million, with an annual debt charge of £20 million. This saves £53 million as a capital debt and £4 million less annual costs. The promoter may also wish to implement the scheme quickly so that revenue can be generated early to repay the debt more promptly and thus reduce accruing interest. These examples (Tables 6.6, 6.7 and 6.8) show the effect of shortening the construction period. The tables below show the impact of interest rate changes.

The project with a 9% annual interest rate, set out in Table 6.8 comes to a total capital debt of £308 million and an annual debt charge of £28 million. The annual cost is higher than either of the options in Tables 6.6 and 6.7.

Table 6.9 shows that with interest the total capital debt is £265 million and an annual debt charge only £19 million.

Table 6.8 Five-year construction period and 9% interest per year

Year	Costs: Infrastructure	Vehicles
1	40	0
2	40	10
3	40	10
4	40	10
5	40	10
TOTAL	**200**	**40**

Table 6.9 Two-year construction period and 7% interest pa

Year	Costs: Infrastructure	Vehicles
1	100	20
2	100	20
TOTAL	**200**	**40**

Comparing the first and second examples (Tables 6.6 and 6.7) shows that shortening the construction period reduces debt charge by £4 million per year, a 21% cost reduction. This gives the light rail operator either increased profitability, or a longer period in which to build up patronage and, hence, revenue to reach the target operating surplus.

The second and third examples (Tables 6.7 and 6.8) show how sensitive the project debt is to changes of interest charges. Raising the interest rate from 7% to 9% increases the annual debt payment by £9 million per year, a 47% cost increase.

The second and fourth examples (Tables 6.7 and 6.9) show that a rapid construction of only two years reduces the annual debt charge from £19 million to £15.5 million (23%) and the overall debt from £274 million to £222 million. All four options are compared in Table 6.10.

Table 6.10 Comparison of the investment options for £200 million project

Table No.	Number years	% interest	Total debt (£m)	Annual repayments
6.6	10	7	322	23
6.7	5	7	292	20
6.8	5	9	308	28
6.9	2	7	265	19

A private light rail promoter should therefore try as far as possible to fund the project by shares, rather than debt. If loan capital has to be raised, then borrowing at a low interest rate and completing the project in the shortest possible time are the best options for achieving a financially viable project early. On completion, the debt can be refinanced at a lower rate, to reflect the asset created and lower risk associated with an operating, cash generating system. In the last example (Table 6.9), if the project is refinanced at 5% per year, then the annual debt interest payment falls from £19 million to £13 million. This is £6 million less, a further saving of 32%.

6.7: Economic appraisal and benefit-cost analysis (BCA)

6.7.1 Affordability

Affordability has a number of interpretations. Here, the meaning relates to a system that can be built with available funding, whether public or private, and the operating revenue exceeds the operating costs. The first part of this analysis is that the tram line does not need an operating subsidy, since this means, for a publicly funded system, it can be operated with only the initial capital grants. This assumes that there is an allowance for asset repair, maintenance and replacement in due course. In a commercially funded system, the operating profit must also service the capital debt. San Diego (Figure 6.10) is an example of an affordable system. The first line was built with only local funding.

Figure 6.10 San Diego, a low-cost starter system – Down Town running

6.7.2 Justification

There are several ways in which new light rail systems can be justified.

(a) Benefit-cost analysis for a publicly funded scheme, where the public and communal benefits exceed the capital costs by an agreed ratio great than 1.0. These benefits will include travel time savings, accident reductions, air pollution improvements and so on, as well as the financial returns. Many of these benefits have no monetary transactions and the values assigned are often subjective and therefore open for debate.

(*b*) Private investors will consider the internal rate of return (IRR). The operating surplus must be big enough to service the investment capital for the privately funded scheme and shareholders receive a fair dividend for taking the initial investment risks. The IRR is determined by the percentage from dividing the operating surplus by the capital cost. The IRR should be greater than the current interest rate. This means that investors will earn more from the tramway project, rather than depositing funds in a bank.

6.7.3 Opportunity cost

When investment capital is in short supply, as it normally is, an important economic concept of 'opportunity cost' comes into play. If the capital was invested in another project, would it generate a higher return? To answer this question two methodologies have been developed. The one in the private sector is older and based on discounted cash flows (DCF) over the life of the project to see if there is a positive cash return, a profit. On top of this there is a second test of the IRR.

This cash accounting method has been adapted by public sector economists as BCA. This includes non-cash items, like the environment, safety, travel time savings and so on.

6.7.4 Discounted cash flow

The accountant's shorthand for opportunity cost is the discount rate. This can be considered as the cost of borrowing money. It is the way in which the private sector assesses how else funds can be invested. The logic behind DCF is that there is a higher utility in having benefits today and paying later. This is similar to personal credit cards.

The private sector gets less benefit from income later in the life of the project, when of course there is more uncertainty. Usually DCF is undertaken for a ten-year projection period in the private sector. Benefits after that are likely to be overtaken by events, such as the changes as experienced in 2008/2009, and include market volatility, credit drying up, new competitor products and so on. A new investment may be needed just to keep the company's market share. If an investment project cannot be justified within ten years, it is unlikely to the undertaken.

This creates a problem for investments with long lives, like light rail systems, power supply networks, generating capacity and so on. The DCF philosophy can still be applied, provided that the investment should pay for itself in ten years. If it lasts for 30 plus years, then it will still be profitable, only reduced by the need for increasing maintenance.

A digression here is useful, to West Germany in the 1980s. Companies were examining the first wave of serious automation to improve labour productivity. Such investments were made, even if the return was no better than the status quo. The logic for that was a tight labour market. Companies found recruitment and retention of staff difficult and were paying staff increasing wages above the level of inflation. The argument was that, as labour costs continued to grow faster than other costs, a capital investment would prove better and need less staff (recruitment and training) cost.

In DCF, annual incomes and expenditures are discounted to the present day by the discount rate, to produce the net present value (NPV). The further into the future these occur, the less will be their NPV. Shortening the time can also have a significant impact on the NPV. The worked example below illustrates these points where an infrastructure investment of £200 million is matched by vehicle supply cost of £40 million, with a 7% discount rate. In this example (Table 6.11), the construction is spread over ten years, which can be compared to Table 6.6. In the second example (Table 6.12) construction is over five years, while in the third (Table 6.13) in only two years.

In this example the NPV revenue/NPV cost = 24.9/175 = 0.14 or 14%.

Table 6.11 Projected DCF with ten-year construction period

Year	NPV Cost Infrastructure	NPV Cost Vehicles	NPV Revenue
1	20	0	0
2	18.6	0	0
3	17.3	8.7	0
4	16.1	0	2.0
5	15.0	0	1.9
6	14.0	7.0	3.5
7	13.0	0	3.3
8	12.1	6.1	4.6
9	11.3	0	4.6
10	10.5	5.3	5.3
TOTAL	**148**	**27.1**	**24.9**

Table 6.12 Projected DCF with five-year construction period

Year	NPV Cost Infrastructure	NPV Cost Vehicles	NPV Revenue
1	40	0	0
2	37.2	9.3	2.0
3	34.6	8.5	3.7
4	32.2	8.0	5.2
5	29.9	7.5	6.4
6	0	0	6.0
7	0	0	5.5
8	0	0	5.2
9	0	0	4.8
10	0	0	4.5
TOTAL	**174**	**33.3**	**43.3**

So NPV revenue/NPV cost = 43.2/207.2 = 0.20 or 20%.

Table 6.13 DCF with a two-year construction period

Year	NPV Cost Infrastructure	NPV Cost Vehicles	NPV Revenue
1	100	20	2.0
2	93.0	18.6	4.5
3	0	0	5.9
4	0	0	6.0
5	0	0	6.4
6	0	0	6.0
7	0	0	5.5
8	0	0	5.2
9	0	0	4.8
10	0	0	4.5
TOTAL	**193.0**	**38.6**	**50.8**

So NPV rvenue/NPV cost = 50.8/231.6 = 0.22 or 22%

Table 6.14 Comparison of the NPV (£ million) of the light rail investments

Table	Build years	NPV Infrastructure	NPV Vehicles	NPV Revenue	Return %
6.11	10	148	27.1	24.9	14
6.12	5	174	33.3	43.3	20
6.13	2	193	38.6	50.8	22

These three examples are compared in Table 6.14 and show the philosophy of discounted cash flow that completing a light rail project quickly and getting into passenger revenue service early is the most viable option, but needs annual capital expenditure to increase from £20 million per year to £100 million.

For any real project, many more options are likely to be evaluated. If too many options are considered there is a danger of 'paralysis by analysis'. These techniques must complement the experience and judgement of a good team, who will confirm their judgement with DCF calculations.

6.7.5 Benefit-cost analysis

In the public sector, where capital funds from taxes are given as a grant, there is a danger that this 'free money' might be wasted, so government finance ministries use BCA to evaluate different projects, not all in the same area. BCA uses a similar methodology as the private sector DCF but includes impacts which have costs or benefits not captured as cash transactions directly by the light rail system. These benefits therefore may be subjective. What BCA cannot evaluate is that the project has been properly specified and designed.

Sometimes these impacts can have indirect financial implications – for example improving safety in transport will mean less medical cost, perhaps even costs for maintaining people who are disabled after a transport crash. There may be other benefits like fewer lost working days. None of these produces any cash income for the transport project but will clearly benefit the welfare budget of the government or other public agency, improve the productivity of work places and benefit the community. So economists create proxy (money) values for these costs and benefits, which are then included in the BCA.

Similarly, a new light rail project may result in the rise in values of property adjacent to the line. This clearly benefits the property owners and indirectly the public sector through an increase in property taxes. Attempts to capture this increase in land values have been made by the use of – for example, 'value recapture taxes', with mixed results.

Where the impacts have a cash effect outside the project, then evaluating them is fairly easy. The difficulty arises when impacts do not have a cash transaction – for example, damaging the habitat of an endangered species that may become extinct. Here there are two approaches. The first is the cost of doing something (different) to avoid the anticipated negative impact. The second is more subjective and needs market research to find how much people would be willing to spend to avoid the impact – for example, by moving house to have less traffic/noise/pollution.

Similarly, some of the cashless benefits might fall on the users of the project – for example, saving travel time. The users are not asked to pay directly for these benefits, unlike a commercial project, where the users pay for the service they enjoy. How can these benefits be evaluated and included in the BCA?

Two basic approaches have been adopted. In the first instance, an attitudinal approach is used, where people are asked how much they would be prepared to pay – for example, to save time. Conversely, they can be asked how much longer the journey time can be, if it reduces their travel costs.

On one hand, the first is easier to undertake but will need behavioural information to validate. In either case the two dimensions of time and money can be plotted on a utility curve, giving a range of 'value of time' figures (Figure 6.11). For simplicity, the average is normally used. From considerable research, there are big differences in 'value of time' between income groups and journey purposes.

The second is a behavioural approach, studying peoples' travel choices and seeing what options, when available, they choose, either to save time or money. The second is harder but gives more credible results.

The work journey, being economically important, has usually been valued, from the kind of research outlined above, at about the wage level of the traveller. This goes some way to explain why the private car dominates the journey to work market, even though car travel is more costly than any form of public transport, including taxis. Other journeys, including leisure, are normally lumped together and valued at about 25% of the wage rate, although there are differences for different journey purposes.

In many transport projects, the main benefit (about 70%) evaluated for BCA comes from travel time savings. As most transport projects are dominated by local trips, the time savings are going to be at most a few minutes. This raises questions for the utility of BCA. Firstly, can such small time savings be detected, especially when travel time variability is large – for example, through

Figure 6.11 Value of work travel time (US 2018)

Source: The author and US Bureau of Transportation Statistics

congestion. Secondly, if detectable, can the time be economically used? In many EU countries, BCA only includes time savings greater than 15 min, which can have an economic use and hence a real value.

When a large number of very small time savings can be used to justify spending tens of millions of public money on a transport project, the BCA becomes very sensitive to the capital cost assumptions. Even at the most confident of capital cost estimates, BCA can show many schemes to be barely viable, especially when depending on travel time savings as justification.

Therefore, it only takes a small capital cost over-run to make the BCA negative. By this time, the project may be too far advanced to stop. Hence the severe criticism of the UK National Audit Office in 2004 about the publicly promoted and funded light rail projects that failed to generate the number of passengers predicted, and so the time savings and revenue. Capital cost overruns have often resulted in corner cutting, sometimes leaving a poor quality or an incomplete system, needing continuing repair or replacement.

6.8. Implementation and phasing

6.8.1 Implementation

The problems identified above can be minimised or avoided if the project manager is confident in the plans prepared and cost estimates for the scheme. The project manager must also ensure that the phasing is such that time overruns do not turn into cost overruns. The key to a smooth and successful new project is confidence and trust among all those in construction, installation, commissioning and operating the new light rail line. This will need continuous monitoring and co-ordination between the different groups involved in implementing the project.

The project manager representing the promoters must be confident that the contractors can do the job on the timescale agreed, quality required and within budget. The contractors must be able to trust the project manager that the contract details will not be changed (significantly) during building and that the payments will be honoured in the stages agreed. This is important because most of the contractors that fail do so through a cash flow crisis, not for lack of profitability.

In the EU and UK, major projects which are funded by the public sector are required to have open and accountable spending. This demands that all such projects (valued over a small cost threshold) are advertised and that a system of competitive tendering is employed. One of the strategies employed by contractors is to price a project very keenly and depend on claims for contract variations to generate the profit required.

This was another area where the UK National Audit Office in its 2004 Report criticised new light rail schemes in Britain. Indeed, a logic for the adoption of design, build, operate and maintain (DBOM) contracts, also known as 'turnkey' (Figure 6.12), is that the risk appears to be transferred to the private sector contractors, since any cost overruns will eat into their profits. In practice, things have not always turned out like that. Often, the biggest risk in such projects is the public sector promoter changing requirements or specifications.

The UK National Audit Office also criticised that implementing each new UK light rail project used a different contractual arrangement with the private sector suppliers. As an example the first light rail project, Manchester Metrolink, was built by one consortium with a single DBOM contract. The South Yorkshire Supertram was a design and build contract with one contractor. An operator and maintain contract with another. When a different contractual arrangement is used to implement each light rail system, no legal or practical experience can be built up, or improvements realised. In contrast, most roads are built to a standard contract (Figure 6.13), so the promoters and contractors have experience and this rarely creates legal problems in implementation.

Figure 6.12 DBOM contract and links

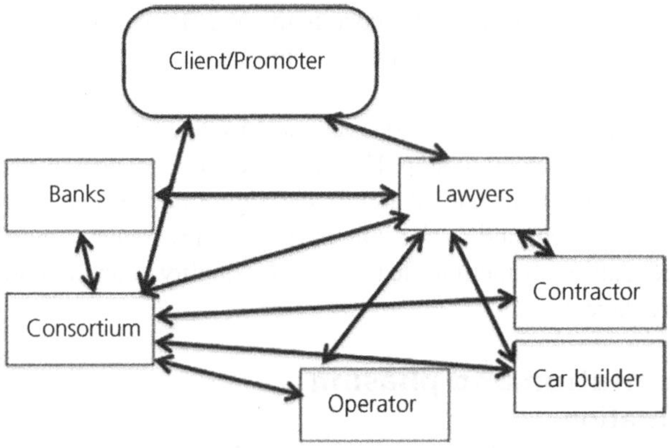

Figure 6.13 Standard tramway contract links

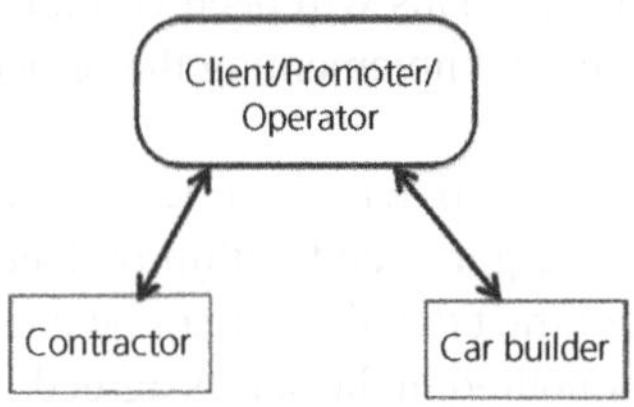

In Britain, each LRT scheme was treated as being new and different, and thus there was very little transfer of learning or experience between successive schemes. From this there were repeated mistakes and cost increases. The light rail schemes reviewed by the UK National Audit Office experienced implementation and interface problems arising from these different contractual arrangements, in which no party had previous experience. Needless to say, this generated a lot of work for lawyers, which the NAO observed took a larger part of the budget than the engineers who designed and supervised the whole system build. In any case, there is a logical anomaly with all these arrangements. This revolves around the assumption that the risk is transferred to the private sector.

In all projects promoted by the public sector, the biggest risk is the public sector itself – for example, change of political control, new regulations and so on. As the private sector cannot possibly anticipate what future public sector risks might be, these have to be factored in by a higher price, as the private sector is also risk averse.

The engineering risks of constructing a light rail or tram system are well understood, with over 100 years of experience worldwide to draw upon. Similarly the risks of operation on the supply side are also well documented. On the demand side, competition from private cars, taxis and so on can be modelled as part of the financial assumptions. There is more difficulty anticipating changes in public policy – for example, fuel duty, parking controls and so on. Factoring in public sector risks, is one of the reasons why, on a comparable basis, UK tramway schemes have cost twice as much as those in EU countries, where conventional client/contractor arrangements are used.

When there is a conventional contract, the public sector accepts the risk from its own decisions and changes – for example, new specifications, location and design of stops, and so on – while the private sector accepts and warrants its contractual responsibilities of workmanship, health and safety to its staff and the general public, and completion on time. The demise of Metronet in London, in August 2007, was primarily due to the public sector changing the specification and then not compensating the private sector contractor adequately, which went bankrupt. Both parties lost out. Indeed the outfall from such collapses is that private contractors have become wary of doing such work for public bodies and fewer bids for projects are now submitted.

Another example of the public sector risks is the high cost to the private sector in bidding for light rail projects. Typically, a contractor will spend at least £2 million bidding for a £300 million, project in the UK, and perhaps win one in five bids. This means over £10 million is spent trying to win a contract, which of course has to be factored into the bid price. Worse still is when projects are cancelled, after contractors have invested so much in submitting bids.

In 2006, public sector tramway projects in Leeds, Liverpool and Portsmouth were cancelled. One of the results of that was AMEC, a major civil engineering contractor, pulled out of any further public sector light rail work, having spent significant sums on abortive bids. Another casualty of this was Laing which went bankrupt due to the failure to write off unsuccessful bidding costs. The organisation was saved by being merged with O'Rourke and continues to trade as Laing O'Rourke.

There is no doubt that a contract where the promoter specifies what is to be done and the contractor builds it, is the least complex arrangement, as well as being the one in most common use throughout the world. It should be well understood by the promoter and the construction industry. Here the risk of the promoter changing specification is taken by the promoter. While the constructor takes the business and performance risks of completion on time, specification and within budget. These are things under the control of the contractor, and there is no need to build into the bid price any extra cost elements. These construction risks, even variable ground conditions, which can have a big impact on foundation costs, can be factored as a normally expected risk, rather than a 'super risk'.

The big challenge for the contractor installing a new light rail system will be the interfaces with a large number of outside bodies, which are neither the client nor the system user. This is because the rail construction work is spread throughout a city, rather than on one constrained and exclusive or private site. Organisations that will have an interest include the Chamber of Commerce representing traders along the route, who might have their access or trade effected during construction. The utility operators of gas, electricity, water, telecommunications, waste water drainage, highway authorities and so on all will interact with the constructor. Getting a good dialogue and understanding before going onto site is important, for the smooth and problem free installation of the tramway.

One way to achieve this is a consultation forum, where the technical issues can be addressed, especially where satisfying one utility might have impacts on others. To make this work, the contractor needs to have clear technical issues to discuss, which cannot be resolved on a one to one basis with utility companies, otherwise the forum is likely to degenerate into a talking shop, wasting everyone's time and reducing the credibility of the constructor.

6.8.2 Phasing

Getting good relations with the local community is important. The community is going to be disrupted, even if only for a short while, and such relations are the key to smooth implementation. A particular street may need to have access temporarily blocked to some houses during construction activity. With planning, the constructor might offer to pay for the residents to go on holiday during the critical time and a guarantee to provide security for their homes during that time. This can be easier and much better public relations than having to work around and severely disrupt the life of the residents, even if only for short time.

Another aspect of implementation will be getting the system open for passengers. The critical factor here will be the location of the depot, since rail vehicles need to be stabled when out of service. Vehicles also need to be valeted, cleaned and maintained. It may, therefore, make sense to build the section of tramway nearest the depot first, both to enable staff training and commissioning but as importantly to begin the passenger service in a limited way. Operational problems can then be resolved before the full system is opened. This will also generate revenue, which will be important for the operator's cash flow. Almost certainly, constructing tracks through the city centre will take longer, as it is a more sensitive environment. Bringing materials to construction sites will be a logistical problem to resolve. If the light rail is built on a moving front basis, then it might be possible to use rail cars to deliver material to site, from the depot which can be used temporarily for material storage.

On the other hand, material deliveries should be on a 'just in time' basis. So heavy road vehicles deliver directly to the track construction site, without the need for intermediate storage and handling, or needless waiting at the site. Material delivery access routes must be agreed and adjusted in co-operation with the highway authority. If this can be arranged, it is likely to minimise the impact on city traffic and, therefore, should be the one which is most acceptable to residents and businesses along the route.

There will be preparatory works for any rail installation, building on surveys undertaken at the design stage, to ensure that no changes have occurred subsequently – for example, to street furniture. These might not impact on the rail alignment directly but could affect the planned method of construction. The next stage will be diverting traffic to allow contractor's staff to work in a safe environment. Building a light rail line is intrinsically no more complex than laying a major pipe for gas or water, with appropriate procedures.

There may be utilities which are physically in the way of track construction and will need relocating. These should have been identified at the design stage and an agreement made with the utility organisation concerned, both for the new location of the diverted plant and replacement equipment. It may be that some plant can be left in situ but with enhanced access arrangements for maintenance. This can include extra inspection chambers, as well as modifying existing chambers – for example, by making them side entry.

The project contractor can choose to undertake all the construction stages, or subcontract some or all to specialists – for example, track installers, (traction) pole erectors and so on. There will be interface problems within the project, which no matter how carefully a critical path chart has been prepared will inevitably result in hold ups. These may be due to late delivery of materials, or unexpected problems when the road is opened up. Continuous consultation should minimise the resulting effects.

When streets in central Manchester were excavated in 1988 for the Metrolink, large cavities under the road were discovered, where sewers had collapsed and washed away the sub soil. The only thing holding up the road surface was the tram tracks abandoned in 1949 and left in situ, which provided short bridges. The cavities had to be filled in before track construction could begin. Fortunately, ground penetrating radar can now locate such voids before excavation and allow for remedial works in the work schedule, including understreet grouting.

Finishing the construction and getting ready for commissioning will invariably produce a list of things not undertaken correctly, which then must be rectified. A contractor will of course try to ensure that everything possible is done right first time, to minimise the snagging list. This saves time and considerable cost, in some cases enough to make the difference between profit and loss. Where the problem lies with sub contractor's work, there may be interface issues, like cables not placed into the right duct.

The final check with the client will be a walk through the system and also with the licencing or approval authority, which may be needed before any vehicle operation can begin. This approval process varies from country to country. In some countries the operator self-certifies the health and safety compliance of the system. In this case, if something goes wrong, the operator will be held accountable, unless resulting from the negligence of a third party, not involved in the service, can be shown. In other countries an independent or government body will inspect and approve for passenger service, identifying any further works needed for safe operation.

6.9. Revenue operation

Before revenue operations can begin, the system, or the part that is ready for operation, will require its operating and emergency procedures to be proven, in many countries, to a licencing authority. As part of this, the emergency services will be briefed and, where appropriate, training sessions and joint simulated incidents undertaken – for example, in rescuing passengers from a crashed

LRV. The traffic police and highway authority will have approved the traffic management and control systems, including LRV operations and, where installed, traffic signal pre-emptions and priorities. This will ensure that no dangers are created and that the traffic management plan works as expected.

This control exercise will also include emergency plans. These should include a power failure – for example, traffic lights go blank and the LRVs stop running, even if only for a short period. Similarly, emergency plans will be drawn up to cope with access for ambulances and rescue vehicles to get to a scene of an incident, and then evacuate any casualties to hospital. Alternatively, there might be emergency practices for problems not on the light rail, but in and around – for example, an adjacent building catching fire. Additionally, there might be practices with emergency services, should there be an incident involving a tramcar – for example, crash (Figures 5.19 and 5.22), fire and so on.

While this may seem like a lot of work, research (Cuffe, 1989) (Figure 6.14) has shown that transport system controllers cannot improvise during emergencies without sometimes making the situation worse. Drawing up and practicing a variety of emergency plans is therefore key to the preparation for revenue operation. Fortunately, most of the functions on a light rail vehicle are fail safe, so that if there is a medical emergency with the driver, the LRV will stop, although this of course might block a major junction. Another member of staff will however be able to move the LRV to a safer location, while the driver is taken for medical treatment. The probability, however, of this kind of emergency is extremely low, especially with appropriate driver training and regular health checks.

After emergency routines have been developed and practiced, the next stage will be to operate the normal service but without passengers. This usually takes between a week and month, depending whether it is a new system or a new line on an existing system. This practice operation without passengers allows staff to become fully familiar with the service, stopping distances, traffic junctions and maintaining a reliable timetable.

Figure 6.14 Information flows to system controllers

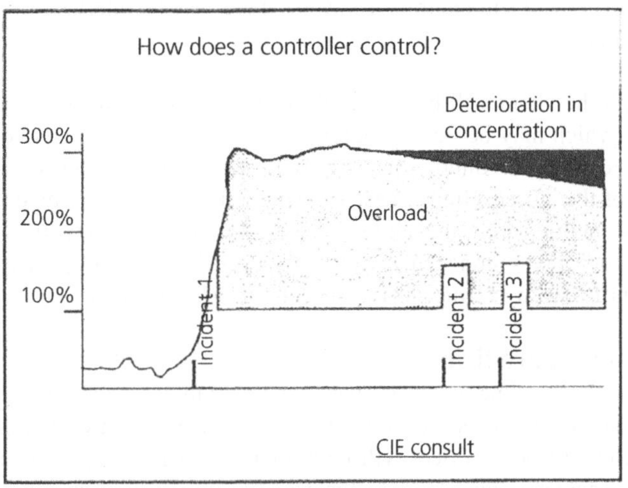

The operator might have groups of residents – for example, elderly and disabled – trying the system out during this period. Most new systems will have at least one day, often during a weekend, when residents will be invited to ride the service for free. With other traffic lower and LRV passenger numbers high, this is a good and final 'dress rehearsal' prior to the start of revenue service the following day.

The aim of a tram service should be to attract the maximum number of people from their cars, to give a safe and reliable service and to help passengers gain confidence in travelling by rail. For the operator the only other aspect of importance is ensuring that passengers pay for their travel. Clearly the fare system used will have a big impact on how this is achieved.

Historically, many urban passengers have paid according to the length of journey undertaken, by graduated fares. In commercial terms this has a large drawback in that it does not fully reflect the cost of providing the service. The marginal cost of running at cruising speed is very low, whereas the cost of stopping and starting will represent at least 70% of the operating costs.

This was realised by American transit operators in the 1920s when the conversion took place from a conductor collecting fares, to one man operation. At the same time a single price ticket (flat fare) system was introduced. This more closely reflects operating costs and also made it easier to ensure fare paying compliance, since passengers either had a ticket or did not, without any grey areas of over-riding.

Similarly, when the labour shortages of the late 1950s made it difficult to recruit tram conductors in West Germany, one man tram operation was introduced with flat fares. In large regional networks, zonal fares were adopted. Compliance was encouraged by having penalty fares typically ten times the flat fare and then inspecting one in ten tickets. On the newly created Rhine Ruhr Transport Cooperative area, (RRVV) in 1978, graduated fares were replaced by system of flat and zonal fares covering the whole are of the Ruhr (VVR, 2026). At the same time, the average fare was increased by 10%, which on the basis of previous experience should have reduced patronage by 3%. Instead patronage increased by 4% and revenue increased by over 14%. This reasons for this success are several, but the resulting surplus was invested in service improvements, creating a positive feedback loop, further increasing patronage.

Getting the fare system right will be an important part of market research for a new tramway. Selling travel by time is another way in which many European systems operate. A ticket giving one hour, four hours, a day, several days or a week of unlimited travel, is a simple way to ensure compliance, and for the passenger the marginal price of extra travel is zero, during the ticket's valid time period. For the operator, the marginal cost of carrying an extra passenger trip is nearly zero, when there is spare capacity off peak. Such fare systems have been aided by the advent of electronic tickets, contactless payments and other methods of automating revenue collection. These help the operator ensure that virtually all fares are collected. It also improves the convenience for passengers, in not having to carry small change or know the fare.

Low cost airlines have revolutionised air travel, with internet booking, cashless transactions and ticket less journeys. In Europe and North America virtually everyone now has a cell/mobile phone. In the UK, there were more mobile phones than people in 2022. Many services can now be obtained by the use of a mobile smart phone. There are, thus, good incentives for new light rail operators to use mobile phones to collect fares and distribute information, as well as collecting data about their passengers.

Once a phone number is registered, a passenger profile can be built up to give information and to sell additional services. These can be in association with activities and attractions in the area. Very few people ride LRVs just for pleasure but are going somewhere to do something. Linking travel with special promotions, entertainment, sports or theatre ticket discounts can be a way to generate extra patronage and revenue, especially off peak when there is spare capacity and marginal costs are low.

Similarly, passengers willing to make a travel commitment some time in advance can be given a lower price than one who turns up at the last minute. This is the policy of many low cost airlines. Indeed it is possible that a simple single journey ticket machine will be on every platform stop, offering a cash ticket at about five times the price of one bought with a cell/mobile phone or bank/credit card.

Monitoring patronage is an important part of maximising revenue. This should be done on a daily basis and the traffic manager must know first thing each morning what traffic was carried the day before. If there is any significant deviation from sales targets, programmes to encourage usage should be implemented. These can be by targeting passengers (with mobile phones), or who live near part of the system, offering special fares and so on. News releases to the local media raise awareness of the service, especially when there is a development. Promotional offers can also be made in the off peak when there is spare capacity. This is an indirect form of passenger/customer feedback.

The operator should also have direct and regular feedback from customers and, as importantly, potential customers. Obviously the operator needs to retain existing passengers as the first priority. Encouraging them to become more frequent riders increases revenue. As important is finding out why potential passengers do not ride. The reasons may be due to perception of the quality of the service, or not knowing the fare system or timetable. Such people can be attracted by public information campaigns, especially those that are directly targeted. The success of a new light rail line will depend on creating a positive public image. In this regard, word of mouth recommendation can be crucial. Added to this, it is as important to have good management by the operator, as important as good system design and construction, and regular maintenance.

6.10. Conclusions

The economics and finances of a new light rail system can be improved in several ways. The most important of which is a good design, implemented right first time. On top of that, the new system can be made more viable by financial engineering. This is important when the investment capital is in the form of a debt, where the interest rate charged can significantly increase the final cost of building and equipping the new system. Probably the most important of these is building the tramway in the shortest possible time. This will reduce the need to service debt before revenue service begins. Starting the revenue service quickly increases the income and the net present value, while the net present debt from construction is reduced, improving the overall financial viability.

LJS Lesley
ISBN 978-1-83708-399-2
https://doi.org/10.1108/978-1-83708-398-520261008

Chapter 7 Marketing

This chapter discusses the ways in which a promoter and operator can communicate with the people who will benefit from a new system, and encourage those who might not, to consider and use the new tramway.

Advertising and marketing must not be confused. These are quite separate activities. Advertising should follow marketing, not the other way round. The purpose of marketing is to find out what (potential) customers want and how much they are willing to pay. The purpose of advertising is to tell (potential) customers what is being offered. Both, however, are aimed at increasing sales and revenue, if it is a product that can be provided economically.

7.1. Marketing

Ideally, market research should precede any decision about an investment in a light rail system. This can seek the best routes, in terms of potential riders and desired traffic attractions. Once committed, further market research will not be aimed at finding out if local residents want a new rail system. It will be about their preferences comparing to cars, buses or taxis for local travel. There are now many ways this information can be gathered, including trips recorded by mobile phones. These can establish and should confirm the travel patterns of residents, and, therefore, existing trips made that could be attracted to a new tramway. Similarly, marketing can focus on such details as colour schemes, brand names, fare systems, service frequency, operating day and customer care.

New light rail system promoters can mimic car manufacturers, transit's principal competitor. The launch of a new motor car is preceded by painstaking feedback from the target customer profiles. This includes workshops into details like the shape, lights, fittings and other things which customers use to differentiate one car from a competitor. Similarly potential light rail passengers can be asked, in a systematic way, their preferences for different aspects of the vehicles, the stops and service. Using indirect methods to assess attitude is likely to be more accurate than direct methods, where the responder may try to give the answer he or she thinks the questioner is seeking. Both of these require the careful framing of questions.

Like car makers, the light rail promoter can also use mock-ups to assess the ergonomics and comfort of passengers who can use the new system. This will also include alternative seating layouts, attractiveness of different seat designs, including coverings and colours. Views about the décor of the inside and exterior of tramcars can be helpful.

Similarly, feedback on navigating a rail system is important, which can be mimicked by virtual reality videos. This will help to assess the information on passengers needs. It can be a salutary lesson for the promoter. Professionals make assumptions about peoples' ability to understand a new travel process for the first time in a strange environment. Such information about the new system needs to be user friendly, which can be tested with a cross section of the local community.

In order to make this exercise realistic, some design work will be needed, for which a cost is involved. Getting it right in public relations and market acceptability terms is important for the financial success of the system. Marketing costs will be much less than the major investment in building the system. An example of this is the West Yorkshire on-line consultation exercise (Figure 7.1) over future public transport investments in the area. Ensuring that responses to such exercises are valid in their own right is also important, especially if the responses are a self-selected sample. Sadly since that exercise was undertaken, despite strong public support, the UK Government withdrew funding for the project.

The science of drafting customer feedback exercises is well established but, of course, the product being tested each time is normally different. Too few light rail systems are customer tested before being built. A preliminary stage is a pilot exercise. This can ensure that the main study results will be valid. The pilot exercise allows wording and understanding to be checked, and, if necessary, the emphasis of the survey rebalanced following the pilot-stage feedback. Spending now can avoid wasting ten times or more later. This will also be much less than the need to retrofit or change equipment, signs and so on after the system has been opened.

Figure 7.1 West Yorkshire on-line consultation exercise

Source: WYPTE

7.2. Advertising and launch

The experience already outlined about the launch of the BEELINE BUZZ Company in Manchester (Figure 5.6), and the reaction of existing operators shows that in the words of Professor Marshall McLuhan, ‘the medium is the message’ (McLuhan, 1967). Often, low-cost direct advertising is more effective in raising (potential) customer awareness than expensive television commercials. Added to this is the use of social media and the internet generally. This is particularly so for travel decisions, when over 70% of all journeys start or end at home. Having valid and attractive information at home will build confidence in the system. Access to information away from home can now be got through mobile and smart phones. An SMS text messaging service should give the normal operation times and, as importantly, out of course service changes and information. This will include traffic accidents and service disruptions. Keeping passengers informed reduces their stress and frustration, and builds further confidence in the rail service.

In 2001 Nokia (Figure 7.2) ran a pan European TV commercial of children playing at home in Helsinki and being alerted by their mobile phones to an approaching tram. They put on their winter clothes, and rushed out just as a tram arrived, when they throw snowballs at it. In Britian the advert was withdrawn after pressure from bus operators who feared it might invite children to throw stones at buses!

Figure 7.2 Nokia tram advert in Helsinki

Source: Nokia Advert

This of course missed a number of points, the most obvious being that no British bus operator, at that time, had a (real-time) text messaging service advising the approach of buses. Perhaps UK bus operators were also concerned that people might associate the 'upmarket' Nokia not with their buses, but with trams. In most of Britain there are no trams. Is any bus operator thinking of converting to trams? Like other consumer products, advertising can be targeted at the appropriate audience. Using market segmentation is both cost effective, and gives good results.

Children can influence parental decisions. Talking to children at school can be difficult but worthwhile. A new light rail system to be built can be preceded by a campaign about the 'dangers' of the tracks. At the same time, positive messages can be given about green, safe, convenient and comfortable travel. McDonalds hamburgers did not get popular by targeting pensioners. Tokens or free tickets can be distributed through schools for families to have a ride, on the Sunday before the start of revenue service. Similarly, parties of school children can be invited to visit the depot to see how the system runs.

Teenagers are the most mobile section of the population, on average each making about 25 journeys per week, some 20% above the population average. Teenagers also have disposable income.

While teenagers often get a bad press for rare outbreaks of bad behaviour, most teenagers are well mannered and active members of their communities. Recruiting teenagers as committed rail riders (Figure 7.3), before they get their driving license, should be important to building a sizeable and permanent market share.

Incentives like frequent rider rewards – for example, T shirts, tickets to concerts and so on – are economic ways to build tram brand loyalty. The most effective way to communicate with teenagers presently must be through their mobile phones. Teasing campaigns were used by the BEELINE BUZZ Company to raise awareness and build market share; not to mention promotional T-shirts given to teenagers.

In most European countries pensioners have free travel or low fares. In the UK, the operators are rewarded not on the number of pensioners carried but on a 'no better no worse' formula. The basis of which is that the operator only gets the same revenue generated by (fewer) pensioners paying the full fare. Many pensioners however have an important family role of baby sitting and child minding.

A light rail operator accepting pensioner concession travel, if compensated by the transport authority, can develop this by offering reduced fares during the off peak for children travelling with pensioners, on the basis that there is spare capacity and the marginal cost of carrying extra passengers is low. Generating extra revenue from accompanied children can be worthwhile, being more than would be the case of near empty vehicles. Children riding with their grandparents are likely to persuade their parents to use the rail system for some journeys.

Figure 7.3 Teenage travel, a survey in London

Children 'prefer buses and trams'

Four out of 10 teenagers are ditching a lift in a car to ride the bus or tram, according to a survey.
20th June 2006
Source: (http://news.bbc.co.uk/1/hi/england/london/5098872.stm)

Pensioners are the least mobile section of the population. Not needing to go to work removes ten of the average 18 trips made per week in the UK. Pensioners over 80 are often effectively housebound. Enabling and encouraging pensioners to make more trips is not only good commercial sense but also performs an important social function. With a fully accessible system, the physical problems of boarding and alighting are greatly reduced. For the light rail operator, many pensioners, and certainly the ones recently retired, may also be car owners and drivers. So the 'treat' of a rail ride with granchildren can be a way to help reduce their dependency on car travel.

Generating such accompanied trips needs the operator to liaise with cinemas, theatres, museums and places where events for children are organised. The tram operator can offer package deals for children and grandparents or other adults to ride and attend the event. The use of internet and mobile phone ticketing makes this easy to administer, with low costs and high marginal revenues, and little fraudulent travel.

Promoting such packages should be undertaken by both parties (the venues can offer 'free' travel with the event ticket). The whole is greater than the sum of the parts in this circumstance, benefitting both businesses. Entertaining children away from the home also has social benefits. School parties can be catered for in the same way, since a class of children can easily be carried in a 200 capacity light rail vehicle, without swamping it. For schools not having to charter buses or coaches is also economic and flexible. The novelty of a tram ride should not be overlooked (Figure 7.2).

Numerically, the largest travel market is adults between 21 and 60, most of whom are also the most committed car users. Here, advertising directed at adults has to address the reasons why people drive instead of using transit. Market research over a number of decades shows the following important high priorities.

- Convenience. A frequent light rail service can offer a turn up and go system, without the chore of having to find a parking space (and paying for it at the other end) or knowing the operating timetable.
- Speed. In urban areas, a light rail system with priorities will offer a faster end to end travel time, especially in the peak, than driving a car. Passengers can use their mobile phones legally on trams and read documents, which is not possible in a car.
- Reliability. Electric trams are the most reliable road vehicles. They never run out of fuel, and rarely have crashes, flat batteries and so on.
- Fares. Usually come 4th or 5th in the importance for decision making to use transit services.

People can only build confidence in a transit system by riding it. This will confirm reliability and the quality of service. Word of mouth recommendation is a most powerful advertising medium. This can be an important way to spread the message. Indeed personal recommendation is often the most cost effective method of advertising.

Some companies employ sales staff to frequent places used by target customers, like bars and clubs, to spread the message by word of mouth. This also is a good way of getting feedback, at least on perceptions. It can provide, therefore, an opportunity to dispel myths. This is a form of active market research. Similarly, complaints can be turned into a marketing tool. Experience on the way complaints are handled shows that positive responses from a company creates a positive and long lasting impact on the complainer and their friends.

The cost of fares are often cited as a reason for not using public transport. In practice, as many surveys have shown driving a car is more expensive but most people do not know exactly how much it costs to drive. Typically 14% of household income is spent on owning and driving cars but the perceived marginal cost of car driving is low. The infrequent filling of the fuel tank is not equated with the journeys made.

Most high-ridership European tramway systems sell access to their services in units of time, an hour, day, week, month and sometimes a year. If coupled with direct bank payment, and the mailing of each new ticket, with promotional material, light rail riding will have an even lower perceived marginal cost than driving. And all extra rail journeys have an effective zero cost to the passenger.

This approach can now be automated, with direct debit payments for travel, and tickets immediately updated on mobile phones.

The most interesting European experience has been in promoting tickets aimed solely at car commuters, so called 'Rainbow Tickets'. In Switzerland, one city (Basel) opted for an economic fare and budgeted on making a slight loss, from the extra capacity needed. In practice nearly 100% more drivers than predicted signed up for the 'Rainbow Ticket' (what's in a name?), so that the operator made an operating surplus, as well as significantly increasing total patronage, reducing peak period motor traffic and reducing the journey time of light rail services.

7.3. Launch and building patronage

In a pessimistic business plan, the operator will assume there will be no fare paying passengers on the first day of revenue service. In the optimistic plan, maybe 50% of target patronage can be expected. Whichever is the starting point, the operator will work on the basis that, initially, revenue will not cover costs. Patronage will, however, increase, as advertising works, experiences are shared by word of mouth and promotional efforts are rewarded. Obviously no light rail system can expect to get a 100% market share but 20% of the trips made in the light rail corridor is a good target. There are examples of European cities where trams get nearly 50% of such corridor trips. If a tramway replaces or supplements an existing bus service, what patronage can be expected compared to the original bus service? Experience in San Diego (Figure 7.4) shows this can be large.

There is considerable evidence on which to base this assessment. Probably the most systematic was the US Transportation Research Board Report No.1221. Here, two different approaches were used to assess the patronage that a light rail service might expect compared to an equivalent level of service with buses. The first set of data came from the replacement of rail and streetcar lines by bus services (Figure 7.5). The second came from the replacement of bus services by light rail (Figure 7.4). Both show that a light rail system can expect to carry between 35% and 44% more passengers than an equivalent bus service, in terms of frequency, stopping places, speed, fares and so on.

More recently, a number of new light rail systems have opened in France, often replacing the busiest bus routes, with the bus network reorganised to feed into suburban tramway termini. These new tramways carry at least 30% more passengers than the bus routes replaced. Interestingly, the bus networks in the same towns also experience an increase in patronage, apparently sharing in the perceived improvement in the quality of the whole network, and of course from extra feeder trips.

Finally in the UK, between 1930 and 1960 virtually the whole of the remaining network of urban tramways were abandoned. In some places trolleybuses replaced the tramways (Figure 7.5), in most cases diesel buses did so. In all cases, the replacement buses were new, sometimes faster and ran more frequently than the 'inflexible' trams replaced. The result was an almost immediate drop in patronage of the order of 30%. Research was needed to determine why but not undertaken at that time. The public sector operators were complacent about their services, and the fact that most passengers were car-less and dependent (but not for long). The abandonment of tramway services became a catalyst for buying cars. This was the same experience in North America ten or more years earlier.

Market research for public services was almost unheard of in those days and indeed, had this not been the case, many tramways would have survived in Britain and France, as they did in Germany, to be modernised later. Why? There was a lot of public support for tramways, and in many UK cities

Figure 7.4 Tram replaces bus service

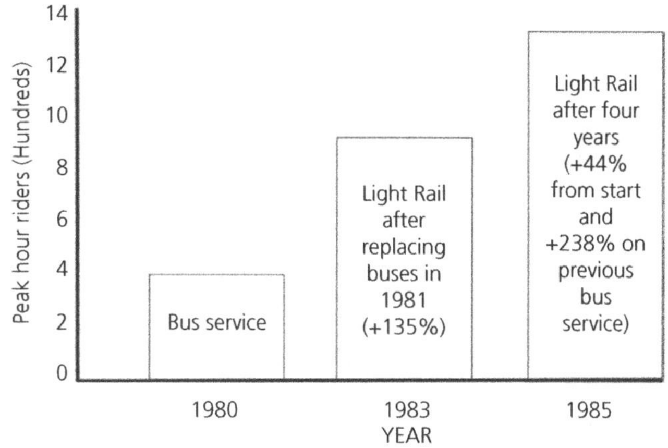

Source: The author

Figure 7.5 Rail to trolleybus in London 1938

Photo: London Transport

petitions were drawn up against the abandonment of tramways. In 1948 in Liverpool this attracted 250 000 signatures, over half of all voters, but was ignored by the city council, and pushed through on the casting vote of the Mayor. Similarly a petition of over 250 000 signatures in 1948 London failed to prevent the abandonment of the remaining tramway network between 1950 and 1952.

On the other hand, today a new light rail system should not be built just to divert bus patronage, since in most places it only has a small and declining market share. The target market must be private car trips. Firstly, because they are the majority in the cities of all developed countries. Secondly, reducing car traffic will create the needed space to allow light rail systems to have priority, both in terms of road space and junction time. This will ensure that new light rail systems remain an attractive alternative. In community terms, reducing air pollution, greenhouse gas emissions, noise and the need to import hydrocarbon fuels cannot be ignored. This should be a crucial part of the justification for new publicly funded systems.

Building new patronage should initially concentrate upon the people living and working in the light rail catchment area. In many places this will be those who can 'walk and ride'. If park and ride stations are also provided, as they should on most lines, then suitable road signage for drivers coming into the city, especially if linked to free flowing lanes into park and ride stations, with congestion on the main road, will attract out of town commuters.

Advanced advertising is certainly a strong part of the McDonald's launch for the opening of all new outlets. Indeed McDonald shamelessly targets children on the basis that their parents will be persuaded to come along, again and again. Capturing customers for light rail at a young age should also be part of a new public transport strategy. An outing by rail will be a new experience for most families, who otherwise go everywhere by car.

To achieve this an 'upmarket' image, as well as a quality service, needs to be presented. Families must be attracted by something better than or at least as good as their car, certainly for travelling around town. Influencing local residents needs a combination of direct information at home, and incentives at the places most often frequented, including work, shops, entertainment and so on. Indeed a free first day, often a Sunday, can be used to get families to try the rail ride, who normally only travel by car. Special attractions in the city could be the incentive to ride trams, even if a joy ride in its own right were not enough. The novelty of a new light rail system should not be under estimated.

7.4. Product life cycle and re-launch

Many consumer products are either periodically relaunched or die (Figure 7.6). A new light rail or tramway system is no exception to this. The question is when is the best time for a re-launch and how should it be done?

An example of this is the South Yorkshire Supertramway, which was opened in 1995. In 1998, it was sold as a loss making concern to the Stagecoach Group, on a 25-year concession. In 2000, Stagecoach made operational changes that included replacing pre-purchase tickets from tram stop machines, with on board conductors collecting fares instead. Stagecoach then undertook a rebranding and relaunch (Figure 7.7). Initially, this concentrated upon a new livery, based on the Stagecoach house style. The original livery was an overall grey, while the new Stagecoach livery is white with orange and blue stripes.

This exercise in 2001 was undertaken after the operational changes had produced a growth of ridership. The re-launch reinforced this, resulting in continuing patronage growth taking the annual traffic from 8 million per year in 1998 to 12 m per year in 2006, and from operating at a loss to being in surplus. Subsequently, vehicle interiors were refurbished and another livery applied from 2007. This signals a new system creating its own positive stamp on the transport market. The 25 LRVs

Figure 7.6 Product life cycle

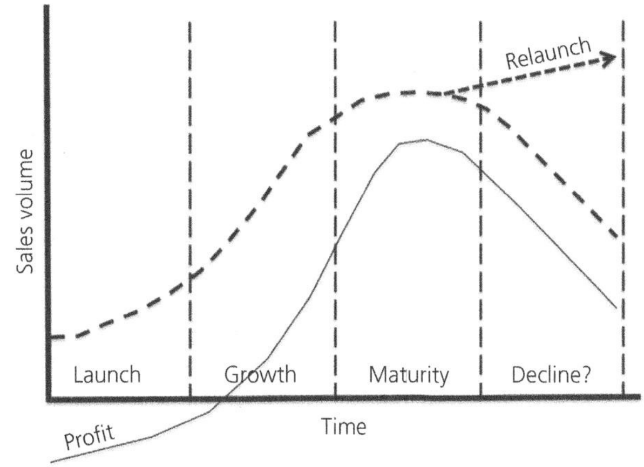

Source: The author

Figure 7.7 Stages of Supertram rebranding 2000–2024

on the 26 km Supertramway, in Sheffield, now carry about 10 million passengers per year, 10% of all transit trips in the city, as bus use has continued to decline.

Figure 7.6 provides some clues as to when and how successfully to re-launch a light rail system.

After opening a new tramway, patronage should grow but, after a few years, reach a plateau. If nothing is done at this point, patronage may begin to decline slowly. Judging the optimum point at which patronage reaches a plateau is the key to a successful re-launch. This will include market research to find out what is good about the service, and what deters non users. Addressing those will be the key to building new market growth.

Stagecoach had strong feedback that passengers did not feel secure in one-man operated tramcars. The introduction of conductors immediately increased the sense of security. It also produced a one off increase in revenue of the order of 7%, as nearly all fares were now collected. This revenue increase, coupled with the savings from not having to repair vandalised ticket machines or collect the money from them, made the new conductors self funding, leading to a profitable period of operation.

Given the timescale for introducing a system-wide re-launch raises the issue of tactics. On a small system with few vehicles, a complete change can be planned and implemented almost at once. On a more complex multi route network, a route by route relaunch will be the most effective, starting with the busiest (most profitable?). Then the extra profits from rising revenue can be used to fund the subsequent route relaunches. On a multi-route system, such relaunches can coincide with the introduction of new vehicles, usually on the busiest route. Vehicles are then cascaded to replace older vehicles on other routes throughout the network. The oldest will then be retired. This at least gives the impression that all routes are getting 'new' vehicles.

First Group has used the brand name 'Overground' and 'Mainline' (Figure 7.8) in relaunching some of its bus routes in Sheffield. This cleverly plays on the reputation of the London Underground for speed, frequency and reliability of service. Additionally, articulated buses were used, initially at half fares in competition with the Supertram Meadowhall Line.

No information has been published on the effectiveness of the re-launching bus services, since the route networks were also changed, going from fine networks with infrequent services, to coarser networks with more frequent services, in order to achieve shorter waiting times. The articulated bus service in competition with Supertram only lasted 6 months before being withdrawn.

Product relaunches can take place for other reasons. A classic was the formation in Germany of the Rhein-Ruhr VV (VRR) in 1978. This is a voluntary association of many public transport operators in the area. The Rhein-Ruhr area is as big as Liverpool and Manchester combined, or Saint Paul-Minneapolis, and with more people. It also had nearly 100 public transport operators. The VRR staff needed to co-ordinate these 100 operators was only 24.

The logic of the VRR was to co-ordinate service information and make travelling by public transport easier. The first major reorganisation in 1979 was a change from graduated, distance route-based fares, to a coarser zonal scheme. This was effectively a fare increase of an average of about

Figure 7.8 Mainline articulated bus in Sheffield

Source: First Group plc

10%. On the basis of previous experience, the budget was based on an elasticity of –0.3, from a loss of 3% of passengers but giving a revenue increase of about 7%. This fare system change was accompanied by good quality and wide coverage advertising, including posters at every station and stop. The result surprised the VRR. Patronage increased by 4%, giving an increase in revenue of 14%. The conclusion from this was that the new fare system was easier to understand and, coupled with the intensive advertising, encouraged extra ridership, including some from non-transit-using residents.

Light rail operators should not underestimate the power of an advertising campaign for a good product but never waste money advertising a bad product. How to know the difference? Ask the staff and local people.

7.5. Staffing

Public transport (transit) is labour intensive. Typically about 60% of light rail operating costs are for staff pay. Given the relatively high cost of staffing, they should be used productively, improving public relations and customer satisfaction. Too often, public transport is run like industrial production, where running the LRVs is seen as the prime objective. Recruiting and training staff is an expensive part of a system start up. Ideally, those employed should stay many years. If staff turnover is high, not only is the cost of replacement expensive but reflects some intrinsic failure to run the operation in a people-friendly way. Many bus companies have to recruit drivers continually, who have a transferable skill. In London the average stay of bus drivers is reported to be less than 12 months. In contrast, when the Tramlink in Croydon in 1999 was recruiting some 85 tram drivers, it received about 1200 applications. This is an example of cost in the other direction.

New light rail must build its patronage by attracting car users. For this to be successful, the service quality is most important. As the Supertramway in Sheffield discovered, introducing conductors can be effective and cost neutral in improving the passenger experience. With an improved travel environment, more patronage will be attracted, especially by the perception of a friendly service and safer journeys. Even the fully automated London Docklands Light Railway had 'Train Captains' on every train. These not only checked tickets and collected fares, they but also provided information to tourists and maintained a good secure atmosphere.

When the bus industry in the UK was deregulated and privatised, the established operators used mostly double-deck, one-man operated buses, where the drivers were paid (in 1986) about £3.50 per hour. Some new operators saw the potential of mini buses to reduce costs.

These new companies did not recruit existing bus drivers, but sales assistants from high street retailers on £1.80 per hour. By offering driver training and a wage of £2.50 per hour, the new mini bus operators were able to recruit staff, already with good customer care. The technical skills of driving a bus are fairly easily learned, especially for those who already have a driving licence. Customer care is a difficult skill to learn.

In public transport services like a light rail system passengers meet staff who, as safe operators, are the system salesmen, the PR office, the information service and so on. These extra and important skills are not always fully included in staff training programmes. Inspectors are also front line staff, even if called 'revenue protection officers' or a similar title.

These staff must conduct themselves well and politely. This can make or break the image of the service. Some years ago, the City of Brussels found that their inspectors, because of the military style uniform, were being called the 'Gestapo' by local people. An authoritarian modus operandi also did not help. This coincided with declining patronage. Which was the a cause and which the effect?

After a rethink, the introduction of new 'softer' uniforms and retraining occurred, along with a poster campaign showing the photo of every inspector, with captions like 'Uncle John', 'Grandad Peter', 'Aunt Mary' and so on, emphasising passenger security and the assistance that was offered. Within six months the public perception of inspectors, and the system, had improved and patronage began to grow. This is another example of an accidental but successful relaunch of a public transport service, which needed good market research.

7.6. Community involvement

Local public transport systems are part of the community and urban scene. In many towns, where multiple branches of large companies line the streets, often it is only the livery of the local public transport service that provides a unique identity to the town. Often such transit operators were founded by local people, further reinforcing the civic link. Until recently in the UK, many were also municipally operated and considered to be an essential service.

An example of how not to do it came from the creation of the National Bus Company (NBC) in England. In 1968 all the state owned and nationalised regional private bus operators were absorbed into a single state-owned company. Many of these bus companies were originally founded by an entrepreneur, and were family owned, reflecting the local communities in which they operated.

NBC decided that a corporate image was needed and, after advice from consultants, adopted a reflected 'N' an arrow logo (Figure 7.9), giving a symbolic impression of movement. NBC also decided that all buses should be either green or red, since many companies already had (different shades of) green or red. There were, however, many others with different livery colours. The destruction of many unique and local brands was an expensive mistake for NBC. By comparison, the Greyhound brand name and logo (Figure 7.9) grew a nationwide bus system, almost from the beginning.

NBC destroyed the loyal identities of many bus companies. This led the author to write an article (Lesley, 1974a) in the trade press, pointing out that many town centres had the same range of multiple shops and it was becoming difficult to tell urban centres apart. One of the few symbols of civic identity came from the livery of the local bus company. With NBC's new standardised corporate

Figure 7.9 NBC and Greyhound logos compared

Source: NBC and Greyhound

style, that disappeared, and the downward trend in patronage of about 1.5% per year that began in 1956, accelerated to about 2% per year.

Today in most towns, very few residents would be able to give a visitor directions to reach a particular place by transit, since few people use bus or other public transport services. Adults mostly drive to work. In Greater Manchester over a third of bus trips are made by pensioners with free travel. Even in London, the most public transport-orientated city in Britain, only about 30% of trips were made by public transport in 2022 (Figure 7.10), whereas 70% were by other modes including cars. This was an improvement since 2001 (Table 7.1) were car was over half of all trips. In the UK, only in central London does public transport carry more trips than are made by private car. London buses also carry over 50% of all bus trips made in Britain.

Table 7.1 also shows that about 1.7 trips a day were made on average by residents of London, and that the maximum public transport use is in Central and Inner London, but even there were only a fifth of all trips. In Outer London, modal split is almost identical to other UK cities, with over half of all trips by car.

In many places, the involvement of the community as transit service users has declined, and with it aspects of involvement and civic identity. Many people do not know who runs the bus or transit service. By contrast, most people will identify with their local major sports team, even if they rarely or never attend matches – supporting 'their' team is now one of the few remaining local identifiable totems.

Garnering local support for public transport is going to be a challenge for a new light rail promoter. Without that, measures of highway priority, either in terms of exclusive use of road space, or pre-emption of traffic signals, might be difficult to achieve. A new light rail system therefore needs a

Figure 7.10 London modal split in 2022

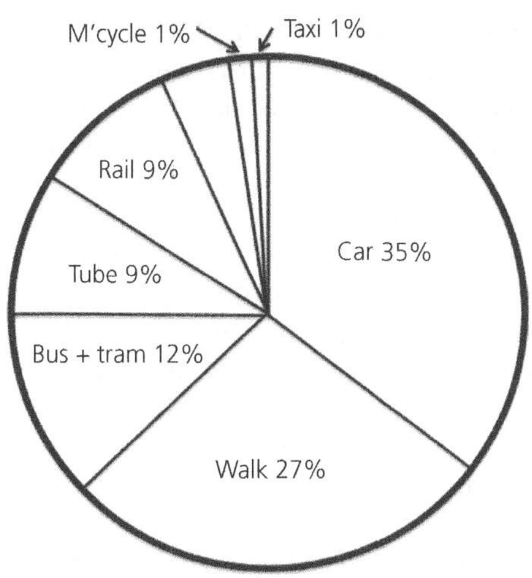

Source: The author

Table 7.1 Modal split (%) in parts of London (2001)

London	Transit	Car	Walk/cycle	Trips/day
Central	20	5	75	0.8 million
Inner	20	30	50	4.1 million
Outer	15	55	30	8.6 million
Overall	17	44	38	13.5 million

well-planned campaign to generate local support, both in terms of it being considered a good idea and to get people on side as potential customers.

When residents (and electors) see light rail as a facility that will improve their area and lives, and give better travel options, then it is likely to win community support. Some of the large supermarket companies are good at this. Today, they often face hostility from residents' groups when seeking to open new branches. Sometimes the reason is the fear that local and familiar district shopping centres will be damaged. Often, a school playing field is sold, built over and children have nowhere for sports. If people cannot walk to such shopping centres, they need a car, unless a new light rail system is integral to the new development.

A new light rail system promoter can form alliances with shopping centres to offer direct and fast rides from the main centres of population in their catchments, allowing some of the customer car parking to be converted into other commercial activities or housing. This is a win-win offer, since the light rail operator can guarantee to capture some shopping trips presently made by car.

These are some of the themes that a new light rail promoter can consider in a campaign to get community involvement, which might also include gaining subscribers as investors as well as riders.

7.7. Route structure and service frequency

There is an art and a science to designing a light rail system, or any transport network. The starting point must always be the objectives that the network is trying to achieve. Where do people live? Where do they make their regular trips? What other locations can attract trips?

For passengers, easy access, short waiting times and fast journeys are important if trips are to be attracted from private cars, which are instantly available and presently the main mode of travel. Service attributes are the most important aspects of a (new) light rail system. The fare is usually a fourth or fifth consideration after the quality of service. The new tram fare is likely to be much less than the cost of car trips, even on a marginal basis. The new operator should be able to offer a fare structure and methods of payment that are acceptable to most of the target customers. These are demand-side objectives and may include direct debit bank payment for pre-purchased tickets or 'contactless' payment for immediate trips.

Operator objectives must include asset productivity. The highest possible operating speed helps, and gives passengers fast journeys. Normally, vehicle productivity is maximised by having long routes, with the terminal dwell or layover time minimised. In a typical city, this will mean routes right across the built up area through the city centre, or around the inner core. This needs a minimum number of vehicles and crew for peak services. Similarly, multi-door boarding and alighting shortens dwell time at tram stops. The optimum stop spacing minimises the access time. It should increase service speed. Trams have high rates of acceleration. With traffic signal pre-emption, a high operating speed is possible. In comparison to buses with a service speed of typically 16 km/h (10 mph), trams making the same stops can achieve 24 km/h (15 mph). Even in areas with a maximum speed of 20 mph (32 km/h) an operating speed of 24 km/h is possible. Trams have a 50% advantage compared to buses, by reducing operating costs and increasing patronage.

On a single light rail line, the journey options for a passenger are 'few to few'. To maximise the benefits of trams needs a network, with convenient interchange stations (Figure 7.11). These will be between light rail lines and with other transport modes – for example, park and ride, taxi and ride, bike and ride, bus and railway stations. Interchanges give 'few to many' travel options but need good travel information and promotion to make realistic, as well as complementary transit schedules, for no wait transfers.

The more lines which interchange, the more journey options are available. This is network synergy. A network provides a 'many to many' service, especially when linked to ride-at-will tickets (Figure 7.12). This will benefit, and attract, the maximum number of residents in the area. The location and design of interchange stations is therefore critical. It should also be recognisable as an important civic focus. An urban light rail network will normally have high frequency services, of the order of once every five minutes. In this case timed interchange is not so critical. The critical factors are then a short walking distance to reach the next service. A 'cross-platform' connection is the best and most convenient.

Figure 7.11 Strasbourg, interchange between two tram lines

Source: Richard Daniel

Clear signage is vital for interchanging passengers to know where to wait for the next service, particularly so when it is the very first journey made, but also necessary as a reinforcement for successive journeys to ensure that the connecting tram is going in their desired direction.

Interchange is also made easier with through tickets, or any time based tickets available (Figure 7.12), since a second payment, acting as a deterrent, is not needed. This was recognised in the 1930s in the USA with the adoption of free transfers. This does not represent lost revenue for the operator, since a passenger making a journey that can be completed on one line only has to pay once, and a journey using two lines has the same utility to the passenger as that on a single line.

Where a second fare is charged, passengers may opt to stay on the first line, and walk from the nearest stop, or not make the journey by tram at all but drive. That is then a real loss of revenue. With spare capacity, interchanging journeys impose no extra costs on the operator and provide greater utility for passengers..

If interchange journeys tax the existing capacity, the operator has a number of strategies. Extra capacity can be added. Alternative interchange stations can be promoted, encouraging and spreading demand. Secondly, fares can be higher at peak times or lower at off peak. This may price marginal peak period passengers to travel during the off peak. Increasing peak fares was the strategy adopted by the Manchester Metrolink in 2005, after being prevented from increasing the passenger capacity of vehicles by removing some seats to let more passengers stand.

A third alternative, where such overcrowding is regular, is that services can be rescheduled. More trams can run in the peak direction only, including skip–stop services in the other direction. Even some returning to the other terminal nonstop.

Figure 7.12 Examples of time-based tickets; Berlin and Edinburgh

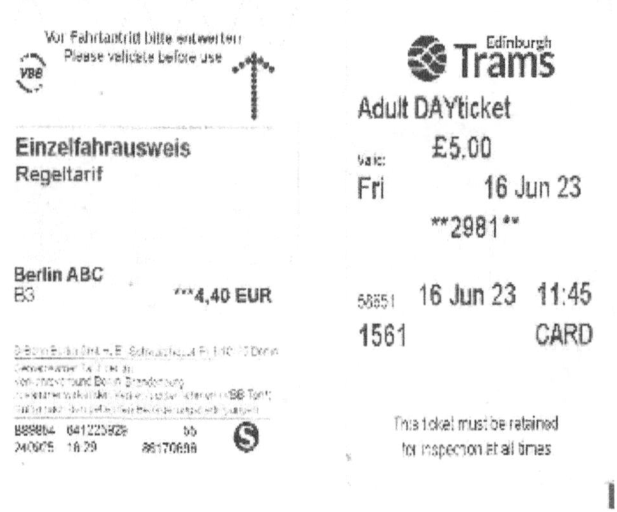

7.8. Conclusions

Marketing and advertising a new light rail line will increase its utility. New lines added after need market research, to determine the location for convenient, purpose-designed interchange stations.. A tramway operator seeking to maximise market share will use regular research to offer an attractive service. Mimicking the convenience of car travel by tram can be promoted with improved speed and avoiding the need for and cost of parking.

The number of routes/lines will depend on the size of the urban area and what can be viably provided. The corridors served will reflect the density and distribution of trips presently made. The service will compete with existing traffic flows. Attracting car trips to a new service demonstrates the success of market research and the effectiveness of advertising. A demand-side analysis must be the first priority. This can be balanced against the supply side cost of installing a tramway. There is no point considering a new light rail route, where there might be big capital costs – for example, needing tunnels, or bridges. Such high costs would be out of balance with the likely benefits (and revenue), and might be a difficult community message.

This is where the use of smart market research is important. This can demonstrate the potential demand and the likely use of a new tramway threaded through the urban fabric. Satisfying the travel needs of residents, with a minimum of environmental impacts, must be part of helping people live more sustainably. This can certainly be part of the promotional message. A new tramway without an acceptable story cannot win hearts and minds, let alone get people to ride.

LJS Lesley
ISBN 978-1-83708-399-2
https://doi.org/10.1108/978-1-83708-398-520261009

Chapter 8
Case studies

This chapter discusses real cities and the introduction of new systems, together with problems faced and solutions adopted.

The preceding chapters have discussed methodologies for the design and implementation of a light rail or tramway system. This chapter considers some recent examples to help show the challenges light rail can face. This includes success stories showing how new systems can be economical, attractive and politically successful. With sustainability and global warming rising up the agenda, a new tramway system can lay the foundations for changing the urban environment, shifting modal split away from car dependency and enriching the travel opportunities for city residents.

8.1. Aarhus, Denmark

The last tramway in Denmark closed in 1972. This first new line was a result of Aarhus City Council deciding to build a tramway and use some existing suburban railway tracks for joint running. This was supported by the national parliament that approved it in 2012, providing 47% of the funding. The city council financed 47% and the balance came from the European Investment Bank, for the rest of the DKK 2.4 billion (about $408 million) project. The author acted as an advisor to Systra, one of the system consultants, and also as a consultant to Danfoss, a Danish electronics company interested in providing equipment for the new trams. The first line construction began in 2013 and opened in 2017. Subsequently, the network has been continuously expanded (Figure 8.1). The network is branded as 'Letbanen', the Danish for light rail. Patronage has grown from 4.8 million in 2019 to 5.8 million in 2023. This equates to 16 trips per capita.

As part of the network runs on railway lines, two vehicle types were obtained – tramcars and tram-trains. The tramcars only run on the city street network (Figures 8.2 and 8.3) at a maximum speed of 80 km/h (50mph), while the tram-trains can run at 100 km/h (62 mph) off street. All vehicles operate at 750 V dc from an OHL system (Figure 8.2).

The environmental impact assessment concluded that there would be an annual saving of 73 t of CO_2 emissions and a reduction of energy consumption of 47 GWh, once the line was operating. Before full electrification was agreed, the option of electro-diesel operation was considered.

The infrastructure was installed by Ansaldo STS and the vehicles were supplied by Stadler Rail. There was one minor problem on the railway alignment, the tram-trains are narrower than railway trains and the platforms edges needed adjustment to enable both types of rolling stock to operate safely (Figures 8.3 and 8.4).

Figure 8.1 System plan (blue line = 100 km/h, red line = 80 km/h)

Source: Aarhus City Council

Figure 8.2 Track in the centre of a dual carriageway road (single wire OHL)

Source: J. Häseler

Figure 8.3 City centre (harbour side) track off road

Source: Aarhus City Council

Figure 8.4 'Railway' station (conventional catenary)

Source: R. Schwandl

The system has been considered a success, so more lines (inter-urban) are being planned. There was a population of nearly 370 000 in Aarhus in 2024. Following the example of Aarhus, a new tramway was opened in Odense, with a population of nearly 210 000. Another tramway planned for Copenhagen was built and opened in October 2025, connected to the six lines of the regional railway network. The new 'LetBane' line is a starter, to replace the 1960s abandoned large tram network, under the policy of 'Overgang til busser'. These will firmly make Denmark a rail-based country and with a large renewable power generating capacity, reduce further CO_2 emissions and make the economy more sustainable.

8.2. Adelaide, Australia

Adelaide is the state capital of South Australia and has a population of nearly 1.4 million. Adelaide is also an example of the impact of the politics on transport policy. Adelaide, like many cities, once had a large tramway but was reduced to just one line in 1958. In the 1990s the controlling Labour Party administration decided to build new tram lines. Before work could start, the control of the city council changed to the Liberal Party, which decided that a tramway was not a good investment but agreed that a mass transit line was needed. After studying options, the choice was made for a kerb guided busway (KGB). The council contracted to adopt a German track system (Figure 8.5), using prefabricated concrete track beams incorporating an edge upstand. As this system could not be installed in city streets, the new line was mostly off street, with access to the city centre by way of existing roads with driver steering and little priority given.

Part of the logic for this decision was that buses could run directly from distant settlements, use the busway and then access the city centre, without passengers needing to change buses. Once open, it soon became clear that any time savings from running on the busway were lost when mixing with existing traffic approaching the city centre. It was also obvious that some settlements did not generate enough patronage to justify a through bus. So there was a need for interchanging at suburban stops on the busway. Added to this, the forecast patronage was not met and the modal shift from car was disappointing.

Then the administration again changed and a Labour council was elected. Pragmatically, the busway was retained but new tram lines were planned to complement the surviving line (Figure 8.6). This has included a complex city centre track junction Figure 8.7).

Making use of some of the wide streets, the new tramway also includes 'lawn tracks' (Figure 8.8) in the city centre. Not only do these give a green aspect but the trams have a traffic free route, enabling faster and more reliable services. The reported cost of these works was about $400 million but the tramway now carried 12.5% of all public transport trips.

Figure 8.5 Off street KGB busway track

Source: Adelaide City Council

Figure 8.6 Suburban off street tramway

Source: Government of South Australia

Figure 8.7 City centre tramway junction under construction

Source: Government of South Australia

Figure 8.8 Green track in the city centre

Source: Government of South Australia

Whether the KGB busway is converted to tramway will depend on how well the new lines perform and the linkages in creating a larger more complex network. It also depends on future councils and if they are willing to develop the tramway, along the proposal for a further five lines, set out in 2008 at an estimated cost of $4 billion.

8.3. Almada, Portugal

Almada is a large town of about 175 000 on the south side of the River Tagus, opposite Lisbon, which has a metro and large tramway which include the steepest (12% gradient) routes in the world. Construction of the first new line in Almada began in 2002 and was opened in 2007. The complete network of three lines, opening in 2008 (Figure 8.9) and meeting near the centre, served hospitals, the ferry terminal to Lisbon, a major station on the local suburban railway and large shopping centres. There is a major viaduct across a valley with the motorway leading to the '25 de Abril' suspension bridge across the Tagos, which has a railway on the lower deck. The system cost of the 13.5 km (8.4 mile) long tramway was €320 million. This gives a unit cost of €24 million per km, close to a European norm. In 2023 it carried 18.1 million passengers, at 1.3 million per route km.

Figure 8.9 Almada tramway network

Source: Metro Sul do Tejo

Some of the tramway is off carriageway in reserved tracks (Figures 8.10 and 8.13). There were examples showing the drainage on these green tracks was not good as envisaged, as mud patches developed.

During a visit to Almada, it was clear that the four-wheel trucks on the Combino Trams and tramway track did not provide the best solution, since there were widespread track failures (Figures 8.11 and 8.12). Some of these were made worse by the interaction of motor cars (Figure 8.11). In some places, the track was grinding the foundation and sand was produced along the track. The vehicles are getting towards a 20-year operating life and no doubt thoughts will be turning to replacements.

It is evident that the tramway is well operated and carries a significant volume of passengers, which compares very well with similar tramways in, for example, Nottingham with 14 million passengers on a 32 km network (0.44 million per km) or Sheffield with 9 million on 29 km (0.31 million per km).

Figure 8.10 Bento Goncalves tramway across a road roundabout

Source: The author

Figure 8.11 Tramway lines 2 and 3 sharing road space

Source: The author

Figure 8.12 Near Ramalha Tram stop with failing track

Source: The author

Figure 8.13 Line 1 tram, curving on reserved grassed track

Source: R. Schwandl @ UrbanRail.Net

8.4. Caen, France

Caen is a city in northern France with a population of about 106 000. Other French cities have built tramways, now totalling 30. Caen City Council thought trams were too expensive for a small city, so opted instead for TVR, a proprietary guided electric bus system (Figure 8.14) which opened in 2002 at a cost of €227 million. The vehicles for the system were provided by Bombardier and the operator was Twisto. The one line had a total length of nearly 16 km (10 miles). The vehicles were single ended and single sided, so needed turning circles at Termini and some intermediate points. From the start there were problems, including the vehicles 'derailing' and, in at least one case, crashing into a traction pole supporting the overhead power supply lines. Obtaining spares for the only TVR in the world was expensive and a continuous problem. This made the service unreliable.

Eventually the council decided in 2011 to abandon TVR. The mayor observed that had they known about the problems, it would have been less costly to build a conventional tramway at the start.

Figure 8.14 TVR guidance system and vehicle on track

Source: Wikipedia and Caen City Council

After the legal issues had been resolved, TVR closed in December 2017 and work began immediately on conversion to a conventional tramway (Figure 8.15). The tramway follows the same route as TVR and a new branch was added together with a new depot. The whole conversion project cost €230 million, about €14 million per km. Given the nearly 20 years between these, the mayor was definitely correct in his comments about the economics.

The new tramway opened in July 2019 (Figure 8.16), taking 18 months to build, also less time than the construction of TVR. Figure 8.17 shows the network configuration, with a main trunk line through the city centre and branches in the suburbs.

Each line has a basic ten minute service frequency, so the trunk section has a tram every 5 min (Figure 8.18). Since opening, the city council has agreed to build a fourth line, to open in 2028 serving the west of Caen. The initial fleet consists of 36 Alstom Citadis trams. A further ten were added in 2024, ready for patronage growth with the new line. In 2022 ridership stood at 16.7 million, and on average residents made 158 trips each. This is a very creditable figure on a tramway only 16.2 km long (1 million per km). It certainly justifies the city council for the extra investment needed to convert to a tramway.

Figure 8.15 Conversion from TVR to tramway

Source: Caen City Council

Figure 8.16 New tram on suburban reservation

Source: Caen City Council

Figure 8.17 Caen Tramway Network

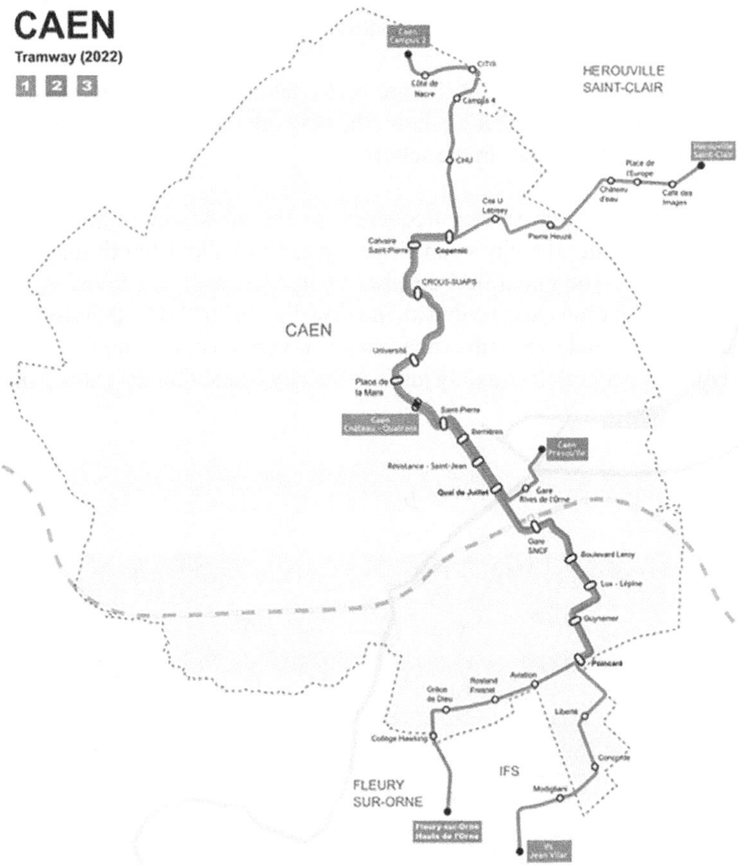

Source: Caen City Council

Figure 8.18 New tram in city centre

Source: Caen City Council

8.5. Edinburgh, UK

Edinburgh is a city of some half million residents, and during the summer, especially during the three-week arts festival in August, there are many more. Its tram project, however, has a number of salutary lessons for anyone considering promoting a new light rail scheme. Edinburgh abandoned its fine tramway in 1956, scrapping recently built modern cars, all double decked and on four-wheel trucks. Neighbouring Glasgow's system continued operating for another six years. Thereafter, Glasgow's suburban railway network was electrified with new 'Blue trains'. Edinburgh made do with just buses, owned and operated by the city council. The idea of a modern tramway was first mooted in 1996 by a private consortium led by PowerGen, a privatised electricity generator. Included in the consortium was the Leith Dock Board looking to redevelop some of the redundant dock estate.

On the advice of the then HM Railway Inspector, the consortium promoted a line from Haymarket Station, along Princes Street, Leith Walk, Great Junction Street, through the Dock Estate to Granton. This was based on a consultancy study undertaken by Scott Hellewell, who had been the operations Director of Manchester Metrolink between 1988 and 1995. Large-scale plans were prepared to obtain statutory approval. This needed the agreement of the city council. The then Convenor of Transport, Coucillor David Begg, refused to agree to a privately funded tramway, offered at no cost to the council.

Instead, he wanted a 'cheaper' busway, the City of Edinburgh Rapid Transit (CERT). This was to be a kerb guided bus (KGB) based scheme from the city centre to the airport. This was costed at over £65 million in 2003, which was too costly for the council; so a shorter version, West Edinburgh Bus System (WEBS), was adopted instead at a cost of only £28 million. This can be compared with the experience of Adelaide's KGB system. Once built, the ride on the busway was so bad, despite contractor repairs, it was abandoned.

There was then a hiatus, until a decision was taken to build a tramway between the airport and Granton, costed at £350 million. Several attempts were made to get contractor bids. Eventually, a consortium led by Bilfinger Berger won and began work in 2007 (Figure 8.19). This included

Figure 8.19 Track construction in Leith Walk

Source: The author

converting the abandoned busway into a tramway. In the absence of any UK contractors, Siemens was appointed to install the tracks. Scott Hellewell and the author met the lead engineer of Siemens in Edinburgh who stated that his company had never installed tramway tracks. None of the work undertaken by the PowerGen consortium was ever considered by the new promoters.

The project was overseen by a 'hands-off' company, 'Transport Initiatives Edinburgh' (TIE). There was a change in the political control of the city council. Minor problems that could have been resolved became an excuse for the contractor to stop work in 2009. There was a gap of some time, during which track laid in Princes Street collapsed (Figure 8.20) under the impacts of buses, before any trams ran. Renegotiating the project led to an increase in cost, initially to £550 million for part

Figure 8.20 Example of failed Princes Street track

Source: The author

of the line between the airport and city centre (12.9 km/8 miles). Cost continued to increase until it reached £770 million on opening the part line in 2016 (red in Figure 8.22).

The project and cost overruns were investigated by Audit Scotland, which reported and echoed many of the points raised by the National Audit Office about English tramways. The principal being that the promoter lacked the necessary skills and experience to specify and supervise a tramway. Spiralling costs were the result of contract changes and extra charges raised by the contractor.

The 27 CAF Urbos 3 trams (Figure 8.21) were delivered by the original contract date and had to be stored idle, until the first part of the line between the airport and city centre opened.

The completed line (blue in Figure 8.22) to Newhaven was opened in 2023 at an additional cost of £210 million. Meanwhile, a public inquiry was set up under Lord Hardie to determine the

Figure 8.21 Edinburgh Tram in Princes Street outside the National Gallery

Source: The author

Figure 8.22 Edinburgh Tramway

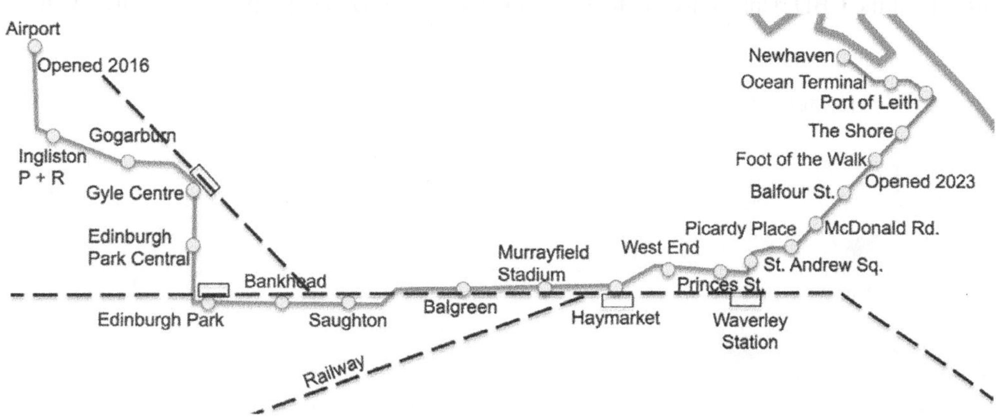

Source: The author

causes of the cost overruns and who was responsible. This reported in 2024 and repeated many of the findings of the Audit Scotland Report. The patronage is now reported to be near the forecast of 5 million per year, or about 0.3 million per km. This, however, is not enough to fully cover the operating costs. The annual report of the operating company showed a loss of £10 million in the last year covered (2024). This loss is part covered by the surplus generated by the council's bus fleet.

TIE confused and compounded the problems of this project, adding a layer of management between the city council (client and promoter) and Bilfinger Berger (Contractor). The biggest risks came from policy changes, that could not be included in the original contract, for which the contractor charged the council extra. Added to this the articulated trams on fixed four-wheel trucks take more time between the city centre and airport, than do buses run by Lothian Transport on congested roads at a lower fare. Edinburgh should be an object lesson in how not to do it. It also highlights the point that there must be proper political and technical structures in place before a project starts. The promoter must also have competent and experienced staff who can specify and supervise the construction of the tramway. Finally, is running a tramway cross subsidised by the local bus service good economics?

8.6. Houston, USA

Houston is the largest metropolis of Texas, with some 2.3 million people. Houston is also a 'freeway' city with approaching 90% of local trips by automobile and only 3% by transit. Like other US cities, its streetcar system closed in 1940, reaching a zenith in 1927. Part of the track right of way was later used for the Gulf Freeway. Between 1985 and 1995 the city created about 16 km (10 miles) of busway (Figure 8.23) a year, to improve transit ridership and reduce car dependency. During that time transit share fell from 3% to 2.7%.

In 1997 the author was in Houston to present a paper at the 'International Energy Week 1997, Structural Dynamic and Vibration Conference'. During this visit he met the Director of Transit at the Metropolitan Transit Authority, who told him about the creation of busways. The data did not show any automobile transfers to bus. Indeed, transit use had declined. When asked if Houston was considering light rail, he was told a line could be opened in 2024.

It was therefore a surprise that the first (Red) line opened in 2004, with a length of 12 km (7.5 miles). This served the CBD, University of Houston campus and the largest hospital in the city. Within a

Figure 8.23 Houston Busway/HOV lanes – with flow and contraflow

Source: Metropolitan Transit Authority

couple of years, this line was carrying 10% of all transit trips in the region. There are now three lines (Red, Green and Purple) (Figure 8.24) with a total length of 36.5 km (22.7 miles), needing a fleet of 104 vehicles to operate the service. These are a mixture of Siemens (65) and CAF (39). The three light rail lines carried 13.9 million passengers in 2023 (about 0.4 million per km). This is 18% of all transit trips in the metropolitan area. It is the tenth most used urban rail system in the USA. Another 39 km (24 miles) of lines are planned. The system has been built to a generous standard with a minimum horizontal curve radius of 107 m (350 ft), allowing a fast service speed, supporting a frequency of up to every 6 min.

This success was after a series of false starts, including proposals for a subway, monorail and heavy rail. All were thwarted by various referenda and changes of political leadership. Eventually, a plan for a light rail system without federal subsidies was approved in 2000. Several lawsuits by residents delayed progress until approved by a referendum. Construction of the Red Line started in 2001, which was opened for service in 2004 and cost $324 million ($27 million per km), funded entirely locally. In 2011 a federal grant of $900 million was given for system extensions.

Rice University has monitored the use of the system. This has shown growing support for light rail and less support for more roads. These have been described as a ‘paradigm shift’ to mass transit. Extensions were opened in 2015 and 2020 (Figure 8.24). More are planned (Figure 8.26).

Figure 8.24 Houston light rail system

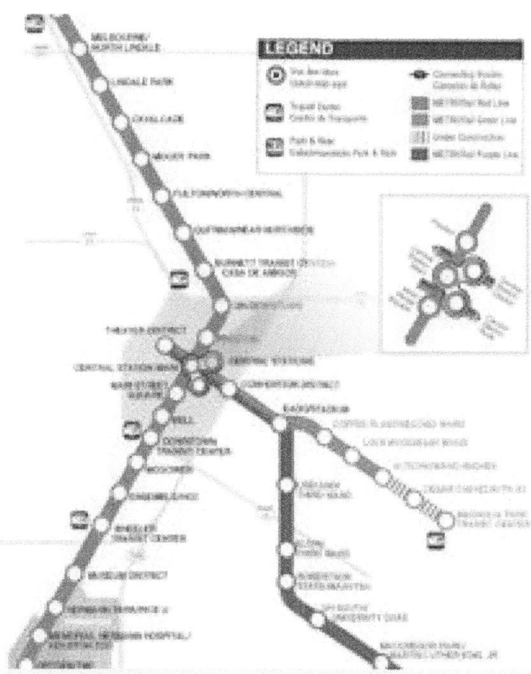

Source: Metropolitan Transit Authority

Despite the modest cost of building the light rail system, the standard of finish is high (Figures 8.25 and 8.27), complementing the city centre and reported to have helped in regenerating some sections.

The change in public opinion towards transit as recorded in the Rice University's monitoring is clearly shown in the style in which the new light rail system has been installed in the Downtown (Figures 8.25 and 8.27). Not only has the infrastructure been sympathetically constructed but the street scene has also been upgraded, including the reduction of highway space available for other traffic. The patronage figures also show the success of the light rail system in attracting riders. Each light rail vehicle carries 130 000 passengers per year, compared to only 50 000 per local bus. In terms of system productivity light rail certainly is doing much better than bus in Houston and follows the experience of San Diego.

Figure 8.25 CBD junction

Source: Metropolitan Transit Authority

Figure 8.26 Light rail expansion plans

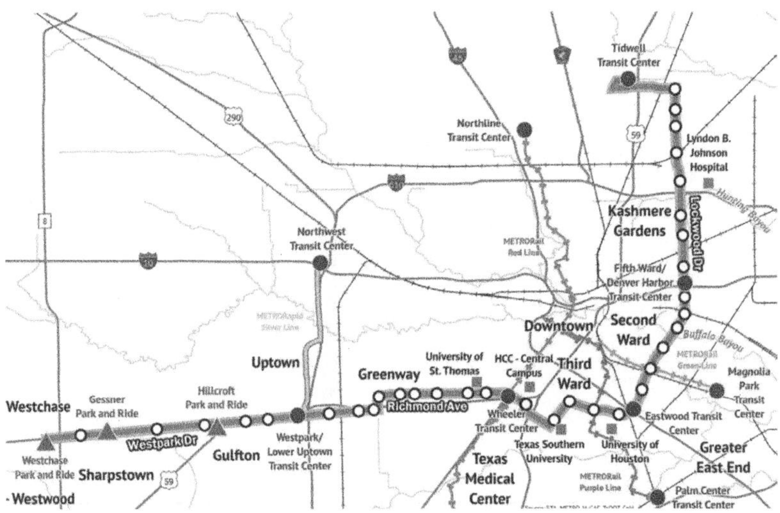

Source: Metropolitan Transit Authority

Figure 8.27 Examples of Houston Street running

Source: Todd Sestero

8.7. Ottawa, Canada

Ottawa is the capital city of Canada but not the largest, with a population of almost one million. This compares to Toronto, 450 km (280 miles) to the west, with a population of nearly three million and a large streetcar/tramway/light rail network. Like many North American cities, automobiles dominate personal mobility. This creates problems of traffic congestion and air pollution. Buses are the main mode of public transport in Ottawa, which has created the world's largest dedicated busway network (Figure 8.28), which opened in 1983. In terms of protecting the 11 routes, which it comprises, from the worst of traffic congestion it has been very successful. In the city centre, however, buses also create their own traffic congestion and add to the high level of air pollution. Some of the network is separate from other roads, and some operates on reserved parts of existing highways. The busway uses a combination of rigid and articulated buses. Despite the relatively

Figure 8.28 Ottawa busway map

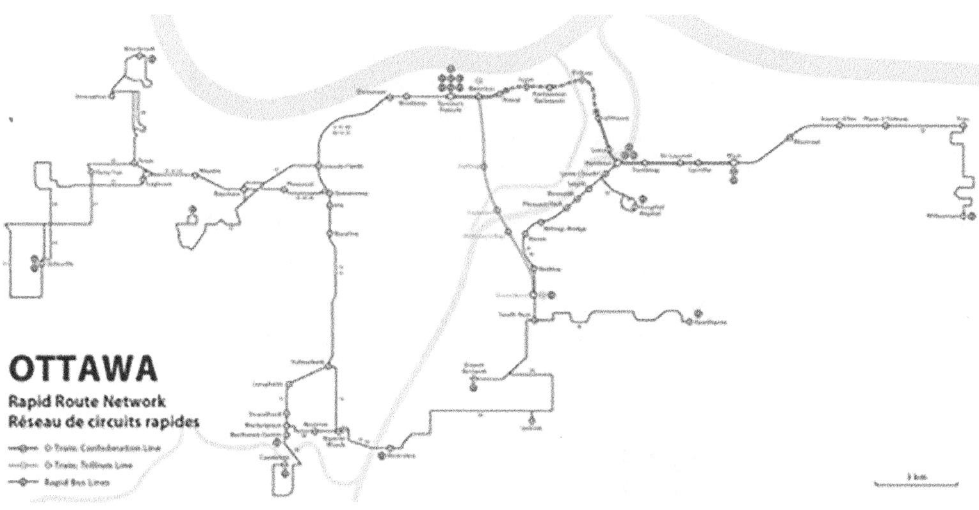

Source: Ottawa City Council

low modal share, compared to European Countries, in North America it is seen as one of the best performing transit systems.

Given the need to improve modal share, which had declined from 14% to 10%, and reduce pollution, after market research, a decision was taken to progressively convert the busway to light rail operation. Market research also shows that trips made per person had been declining, from 1005 per year in 2011 to 910 in 2022, partly as an impact of the Covid pandemic and an increase in working at home. Construction of the new light rail system branded as 'O-Train' began in 2015, converting a busway route to form the 'Confederation Line'. This was completed and opened for service in 2019.

More busway sections were closed in 2021 for conversion to 'O-Train'. The success of this conversion can be seen from the increase in patronage generated. In 2023 nearly 19.5 million passengers were carried just on the Confederation Line. This compares to 13.9 million on the whole of the Houston system (Section 8.6). Like Houston, the 'O-Train' operates at a high average speed of 60 km/h (37 mph) (Figure 8.30), which makes for a very productive use of vehicles and staff.

The whole 'O-Train' network is grade separated from road traffic and in the downtown there is a short subway section, which would not have been possible with diesel buses. All vehicles are low floor with level boarding and alighting from low platforms. The current two lines in operation are being extended, opening in phases until 2027. A new Line 4 will be built to serve the airport. Line 3 will serve the west of the city and is due to open in 2027.

Considerable effort has been made with public consultation (Figure 8.31), with an example of the catchment area of a planned new stop.

When the four lines are completed, the network will be 64.5 km (40 miles) long, serving 41 stations. While this has significant light rail characteristics, being fully separate makes it more like a metro.

Figure 8.29 Converting busway to light rail

Source: Ottawa City Council

Figure 8.30 'O-Train' in highway median strip in snow. Note the check rail

Source: Ottawa City Council

Figure 8.31 Plan of Covent Glen stop

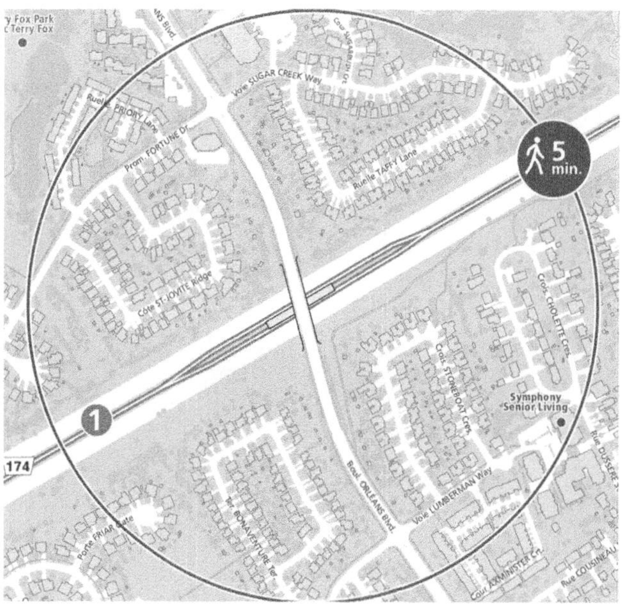

The vehicles, however, are clearly low floor 'trams' and could operate on roads, with suitable traffic management, as occurs in many places including Houston (section 8.6). Should, therefore, Ottawa seek lower cost extension to the O-Train network, street running with priority for trams would be both operationally feasible and bring the system closer to more people, increasing the utility of the network.

8.8. Preston, UK

How large must a city be to support a tramway or light rail system? Preston is a study of an aspiring system. The city has a direct population of nearly 115 000 and in the immediate area nearly 20 000 more. It is the administrative centre for the County of Lancashire. The city is entirely on the north bank of the River Ribble and the conurbation south of the river is in a different local authority area. The local government structure in Preston is, however, divided. Lancashire County Council is the transport authority, whereas Preston City Council is the planning and housing authority. This inevitably leads to some clashes of policy.

Until recently, the county council has followed a road building and car traffic enabling policy. One of the results of this is that Preston has a car dominated transport system, with over 70% of trips, whereas public transport (bus) has less than a 10% market share. Preston City Council on the other hand until 1993 owned and operated local bus services. These began life as a six-line 17 km (10.5 miles) long tramway (Figure 8.32). This took over the original privately owned tramway. The municipal electric tramway opened in 1904 and operated until 1935. Preston was also a centre of tramcar building in Britain, so all the council trams were built locally.

A group of local people began work on promoting a modern tramway. Firstly, a study was undertaken of potential lines and then of their possible financial viability. As a result of this, Line 1 gave the best return (Figure 8.32), followed by Line 2 and Line 3. From this, a household survey of the 18 000 people who live along the proposed Line 1 was undertaken to determine their present travel patterns and willingness to use a new tramway.

Figure 8.32 Preston Tramway Network 1904–1935 and 3 possible lines

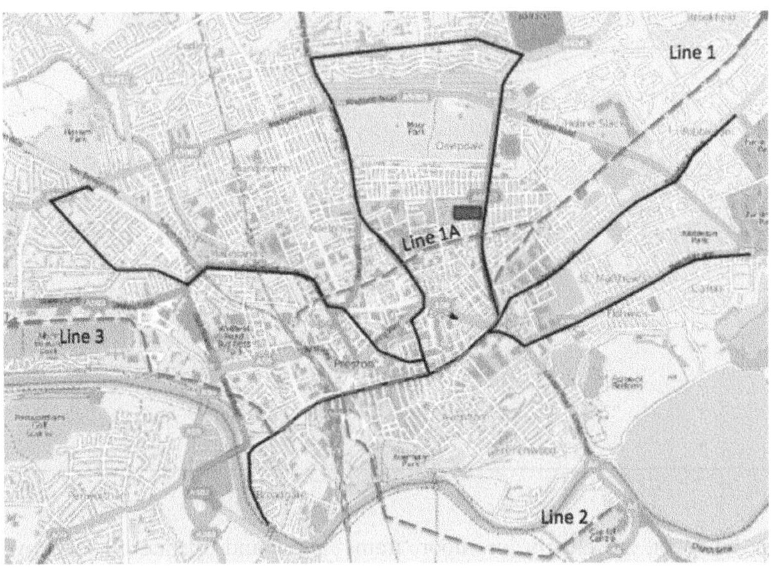

Source: The author

Line 1 is mostly on a former railway alignment, built in the nineteenth century for the movement of freight in and out of the area. It opened in 1838 and operated as a free-standing line until connected to the national network in 1885. Passenger services ceased on the line in 1935 due to competition form the local tramway. Track remains on about a quarter of the alignment, which can be modified (Figures 8.33 and 8.34) for the use of low-floor trams. It also will provide a separate footpath and cycleway.

One of the complications is that the alignment is in two ownerships. About a quarter is owned by a national railway company (Network Rail) and the rest by the city council. The latter section has been converted into a combined footpath/cycleway. Given the generous width of the original railway alignment, a double track tramway and realigned footpath/cycleway are possible (Figure 8.33). As there is little interaction with existing roads, a high service speed of 35 km/h (22 mph) could be possible. Immediately to the south is Ribbleton Lane, where peak hour road traffic speed is typically 15 km/h (10 mph).

An outcome of this work has been to discuss with the local university (UCLAN) which has a large campus and 42 000 staff and students. The main UCLAN campus is at the west end of the old railway line, extending the tramway (Line 1A in blue Figure 8.32). This would allow a tram stop to serve the staff and students, as the campus has limited car parking and cars are presently

Figure 8.33 How the existing railway can be modified

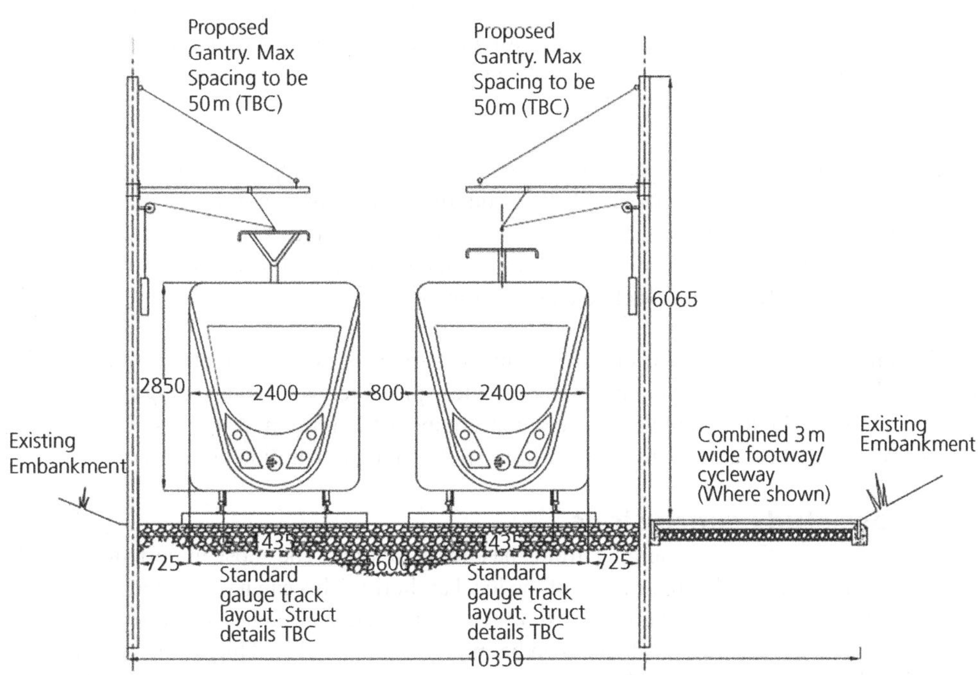

Source: The author

Figure 8.34 Example of a former railway converted to tram operation

Source: The author

the dominant means of travel. The university is trying to discourage the use of cars to access the site, for which a new tramway could be very helpful. Locating a depot to provide stabling, maintenance and repair is being finalised. At the east end, the line meets the outer city bypass (M6), where a park and ride site will be located to attract external trips, additional to those made by local residents who have a high level of car ownership and can walk and ride to one of ten stops on that section.

Local city councillors are firmly behind the project, which has been approved unanimously by the city's planning committee (Faulkner, 2020). The county council however has created several difficulties for progressing the tramway. It remains to be seen if the change of UK national government, committed to encouraging private investment and a change in the political control of Lancashire County Council will alter the approach to a new tramway.

When this initiative is running, a sustainable tramway service is expected to attract about 20% of trips presently made in the corridor, and being privately promoted will be repeating the history of 150 years ago when the first tramway was opened. At about £20 million, it will also be extremely good value for money for a tramway nearly 5 km (3 miles) long (£4 million per km).

8.9. Case study conclusions

There are some clear threads emerging from these case studies. The most important being the impact of local politics and the importance of the local authority. This can have a huge influence on the improvement or otherwise of public transport (transit). There are roughly two approaches. The first is that bus operation is less costly and can be implemented quickly. This ignores however the considerable data and evidence that bus systems do not get people to leave their cars at home, or to change modal split in favour of public transport. Nor do bus systems improve the local environment or level of air pollution.

The studies show that a large population is not vital for a successful light rail line development, provided the route is chosen to maximise the catchment of existing (car) trips. The UITP previously showed (UIPT, 1991) that a population density of about 1000 per km^2 (2600 per sq. mile) is sufficient. Similarly, there needs to be clear transport objectives, which are understood and accepted before any alternatives are examined (UITP, 1991).

The second approach is that the main aim of improving transit services is to reduce car dependency, improve the environment and speed up travel. These objectives are possible with light rail transit. While a subway system is ideal, the cost, disruption and long implementation time makes this unrealistic in most cities. Light rail/tramways provide 90% of the benefits of a subway at 10% of the cost, and can be quick to install. A subway would be inappropriate and uneconomic in a city as small as Preston, a similar size to Caen (Section 8.4) which opted for a tramway.

With appropriate traffic management, a high service speed of about 25 km/h (16 mph) is possible, which will attract a significant portion of trips presently made by private car commuters, removing the need to provide parking space. In evaluating options, all should be on the same basis, with 'do nothing' as the base case. If benefit-cost is the method of analysis, all benefits must include depreciation, discount rates, inflation and economic and population growth.

Another thread emerging from these case studies is that specifying and supervising a new light rail line is the key to ensuring that costs do not run away. In those countries where there has been hiatus in light rail line construction and operation, it makes sense to employ experienced engineers from places with continuing light rail systems, both to avoid costly mistakes and to use best practice. This is especially important when the investment needed comes from a small pot, with competing calls in other social and economic sectors like health, education, defence, housing and so on. The capital costs can be kept low by using existing rights-of-way and only applying construction standards appropriate to light rail.

Operating costs are an on-going commitment which, if not fully covered by patronage revenue, require subsidy; another call on public funding. Operating costs can be reduced by a high operating speed, which also reduces the number of vehicles and staff and hence the initial investment cost.

One object of this book is to bring together both best practice and examples of avoiding things that can go wrong.

LJS Lesley
ISBN 978-1-83708-399-2
https://doi.org/10.1108/978-1-83708-398-520261010

Chapter 9
Conclusions

This chapter draws together conclusions from the book to enable a prospective system promoter to learn from best practice and avoid costly mistakes.

9.1. Background

The art and science of designing and engineering new tramways or light rail systems has been reviewed in this book. A considerable body of experience and research has been drawn upon. In nearly 150 years of electric light rail systems there is experience enough to find cost-effective solutions to most, if not all, problems. These can be applied to the problems of installing a new system in existing urban areas. Much research, especially 'after' studies, shows that a new light rail system can provide many transport, environmental, business and community benefits, including a significant attraction of trips from private cars and reduction of traffic congestion.

Most new tramways and light railway systems are promoted and funded by public bodies using taxes for finance. Historically, most rail systems were built and operated by private companies. Over the years many were taken over by public bodies or closed down. This book has therefore considered both models.

The first is where public authorities use public money to build and open a rail system, hoping to generate wider benefits like reducing traffic congestion and air pollution, and improving accessibility for people with disabilities. Most public light rail or tramway schemes tend to be in regulated markets, with fares set politically, and the level of service determined by the willingness of the city council to fund operations.

In the second, private companies use private funding to build light rail systems. The aim is to produce operating surpluses that can service the investment capital, either by interest payments or dividends to shareholders. The private funding model is most likely to succeed where regulation relates primarily to health and safety matters, with commercial freedom to set fares and the level of service reflecting market conditions.

In a third hybrid option, the public sector forms a deal with private bodies, which raise the funding, like a mortgage, to build, equip and operate a new system. The private consortium has a concession period where the public sector may pay to redeem the investment cost and complement any profit for the private companies. This hybrid has been developed in the UK as either a 'private finance initiative' (PFI) or a 'public-private partnership' (PPP). While this has the benefit of keeping the investment debt from the public financial books, it is also a more costly method than either the first of second options above and spread over a much longer period, as discussed in Chapter 6. (especially Section 6.5).

This raises the question as to why there is a need for new tramways or light rail systems. This was discussed in Chapter 1. As urban areas grow, the demand for travel increases faster than can be accommodated by road building, which is expensive and land hungry. In cities with high levels of car ownership, building enough highway capacity was shown as early as 1963 (Buchanan, 1963) to be impossible (Figure 9.1).

Many US cities have tried road building (Figure 9.2). This merely attracts further and longer car trips, with peak congestion lasting nearly all day. Not only is this self-defeating but also converts land that adds value to the local economy (Strong Towns, Marohn 2020) into becoming economically negative. By contrast, a two-track tramway or light rail route (7.6 m/25 ft wide) has more passenger capacity than a six-lane road used mostly by automobiles (30 m/100 ft wide).

Data from London shows that there was a period when more vehicles were entering the central area but carrying fewer people (Figure 9.1). During the 1960s there was a plan in London to build

Figure 9.1 Regent Street, London 1960

Source: Reddit

Figure 9.2 Atlanta freeway

Source: Curbed Atlanta (CBS News, 2026)

a freeway around the central area about 6 km (4 miles) out (Motorway Box) (London, 2017) in order to 'solve' traffic congestion. The financial cost and disruption this would have caused over several decades led to its abandonment and the adoption, instead of incremental traffic management schemes.

What the 'Motorway Box' might have produced can be seen in the aerial view of a US urban freeway (Figure 9.2) where a large area of land has been converted into a freeway, and other land is used as parking lots. All of which are value-subtracting economic activities.

9.2. Meeting community needs

For an urban economy to operate efficiently, people need access to work, schools, shops, hospitals, leisure and the other facilities. As Section 9.1 shows this cannot be achieved by everyone driving private cars everywhere, even in freeway rich North American cities. Daily examples of serious traffic congestion and air pollution lead to health problems and time wasted.

European cities with high densities of development can only meet a small part of the demand to use private cars. This has been known for many decades, but the political 'supertanker' can take a long time to change course and policy. Often, public opinion is well ahead of the views of city leaders, in realising that there needs to be an alternative to road building and accommodating private car use. Economists have demonstrated that land needed for roads and parking reduces the economic output of the city. There is a critical proportion of such economically 'unproductive' land.

This ignores the energy issues, dependency on fossil fuels, the environmental problems of greenhouse gases leading to global warming and other health threatening pollution. Most developed cities with high levels of car ownership experience toxic smog (Figure 9.3), especially during

Figure 9.3 Traffic smog

Source: plherrera

periods of weather inversions. Such air pollution has health impacts, resulting in hospital admissions and lost days from work. Smog also shortens life expectancy and, therefore, reduces the available workforce.

Access to services means people without cars should be able to walk or cycle for short journeys (in UK 50% of all trips are under 5 km (3 miles) long). For longer trips, public transport, including taxis, are an option. Buses are the only public transport in many cities. Over the last 50 years bus use has fallen in many countries, as car ownership has risen. Car users do not find buses an acceptable alternative. Car users will, however, use a rail service when available.

Many cities have an aspiration for a metro, subway or underground system. These are very costly and slow to construct. Light rail and tramways are equally attractive to car users and provide 90% of the benefits of a metro at about 10% of the cost. There is the added attraction of quicker installation as the case studies in Chapter 8 show.

The advantage of light rail is that stops can be close together, and so is proven to be an acceptable alternative to private car use. Car users will park and ride at a light rail stop. On top of this, electric light rail emits no pollution in city streets, is quiet, comfortable and accessible to all groups of people in the community. With renewable electricity generation, light rail can also be completely energy sustainable. Using renewable power helps to reduce need to burn fossil fuel and so greenhouse gas emissions. This is a double benefit, when diverted car trips save fuel and reduce greenhouse gas emissions are included.

Light rail or tramway stops can be closer together than metro stations, so offering shorter access distances. This means that door to door travel times are reduced. Indeed, a tramway can be faster than a metro for journeys under 8 km length (Figure 2.11). In the UK about 70% of all journeys are less than 8 km long. Furthermore, tramway stops, being at street level, provide easy use, a high degree of personal security and so passenger confidence.

9.3. Satisfying market demands

Urban travel patterns are usually many origins to many destinations. A lot of city centres (CBD) have lost their dominant economic roles. In 1960 some 70% of all trips in Liverpool, UK were to or from the CBD. By 2000 this had fallen to 30%. A new light rail system needs to cater for non-CBD based trips, and from experience of new lines in the USA, also helps to re-establish the role of the CBD by releasing car parking (in Liverpool 70% of CBD work trips are by car, despite a metro system).

Changing the modal split by reducing car dependency for urban trips will enable more intensive economic activities to be developed. These can be accessed by more people using a light rail system and reduce the need for car parking. With the growth of e-trading, there may also be a need for less face-to-face contact. Previous experience shows that better communications stimulate travel.

The growth of e-communications has allowed an even wider group of people to make contact and has reduced the need for some personal meetings. Many are now virtual, over the internet. Indeed, such changing markets can stimulate use of socially acceptable light rail, by raising the awareness of and informing about services available. This is particularly true for young people, who are the

Figure 9.4 New tramway in Lund, Sweden

Source: Lund Tourist Office

most mobile section of the population – for example, university students, who find light rail 'cool'. Many of the new tramways have been built in university towns like Lund (Figure 9.4) to serve at least the main campus. For those brought up in an automobile culture, the switch to light rail is acceptable for many journeys, as the studies by Rice University in Houston is demonstrating. Serving key attractors of trips, like hospitals, universities, even the city centre is a key to attracting passengers and achieving operational viability.

While the way we work and earn our living will change, as will shopping, education, leisure activities and so on, the basic need of human society is personal contact. New light rail systems enable more people to make contact and enjoy social travel.

This means less resource consumption and less environmental damage. New light rail operators will use the internet to determine the travel needs of residents in the area and tell them about the convenience and advantages of the rail services offered. Light rail may become a mostly cashless activity, and so perceived to have a lower marginal cost than even driving a car.

Light rail systems can offer rewards for regular users, as a special treat, or to reach areas not directly served by light rail. Similarly, a light rail ticket could include a taxi link or a hire car. Finally, bike and ride, with cycle trails from light rail stations, is another option for future urban leisure travel already enjoyed by residents in many European cities.

9.4. Getting things done

Knowing the technical requirements for a safe and attractive light rail or tramway system is one thing. Getting one built from scratch is another challenge as in Tunis (Figure 9.5), where second-hand trams from Hannover were first used to reduce the initial capital cost. Not only does the promoter need a team of experienced and qualified people, there is also a need for a public champion. Ideally, the champion should be well known in the area and respected by the politicians and officers who run the city council.

Figure 9.5 Tunis, new and secondhand trams

Source: Metal Traveller

If the project is new and being promoted by the council, then it is unlikely there are already suitably experienced and qualified staff employed. New staff will be needed. Early on, clear objectives must be set down, agreed and stuck to (UITP, 1991). Market research of the local population who will use the new light rail service is essential if it is to be successful. This research needs to find the best route, location for stops and the quality of service that will attract many trips presently made by car.

9.5. Learning from experience

Fortunately, there is considerable published research from earlier light rail projects that can be drawn upon, both for guidance and corroboration. This was used by Bergen (Figure 9.6), a city of nearly 272 000, which called upon the experience of Oslo, where there had been no cessation of service since the nineteenth century.

During project gestation, good relations with local media is important, both to get the message over and to benefit from feedback. While it is always preferable that messages given to the media are

Figure 9.6 Bergen, Norway new tramway

Source: Markus Mainka

positive, sometimes there are difficulties where honesty is definitely the best policy – for example, the problems that beset Edinburgh (Section 8.5) led to political fights.

9.6. Get it right first time

The starting point of a new tramway is route selection. From this is the development of implementation plans, which contractors will use to construct the new system. At each stage it is important that problems are avoided and construction is on the basis of right first time. This avoids the high cost, delay and embarrassment of retro repairs.

Experience has shown that where the right routes have been built, passengers will use a new light rail system, many switching trips from cars. This ensures that staff are fully trained and can operate the service safely and efficiently. This is important to achieving financial viability, where revenue generated exceeds operating costs.

Good market research is fundamental (Section 7.4) to providing an acceptable combination of service quality and fare levels. With the advent of touch-free and cashless payments, the operator should know immediately the income, and the level of honest travel. Should either costs or revenue deviate from plan, then the operator can implement adjustments to avoid finances becoming intractable.

9.7. Diversification

To get people to use a new light rail system needs convenient and attractive stops, which market research can help to optimise. The form and appearance of the vehicles complements the stops, as the public engage with the light rail system. Such vehicles may be articulated to enable operation around small radius track curves as short as 12 m (40 ft) radius, with a large proportion of seats to accommodate all off peak passengers. In the peak there should be enough safe space for some passengers to stand.

While the first new light rail line in a city may be the most difficult to design and build, the capital costs must be controlled. Without an economic first line, future extensions may be difficult to justify. Nonetheless, designing the first line will have an eye on possible extensions or additional lines. A network will have a larger passenger utility and attraction than a single line. Exploiting network synergy can make a good first line into a successful network.

9.8. Building a network

Political support is vital, especially if the light rail project is publicly promoted, since as some examples in Chapter 8 show a change of political leadership can result in much good work being lost. Delivering an attractive new light rail system economically, providing positive value from the investment is important, especially when the patronage revenue covers the operating costs. Additionally, wider social and environmental benefits can be achieved (Figure 9.7). These are a bonus for the community that does not appear in the finances of new tramways.

With a wide appreciation of the need for environmental sustainability, light rail can be a clear demonstration of the commitment to a greener urban transport system. On average, a new light rail line is opened every two weeks worldwide, indicating the perception and acceptance of modern rail transit. Freiburg in Germany has used an expanding tramway network to attract 'green' companies and provide extra accommodation for a growing population, with less use of cars for everyday travel.

Figure 9.7 Freiburg extensions in Reiselfeld and Vauban

Source: Freiburger Verkehrs AG

9.9. Conclusions

New tramways or light rail systems attract trips that are made by automobile, which 'cheaper' bus based systems do not. They offer most of the benefits of a full metro/subway/underground at about a tenth of the cost, and with much less disruption of the urban fabric New systems which initially faced opposition from residents or business, once open quickly lead to other parts of the city lobbying for a line. Lines joined into a network create travel synergy, expanding from few to few, to many to many journey opportunities. Successful systems, e.g., Freiburg, have seen car ownership and usage decline as tramway ridership increased. Patronage translating into revenue that exceeds operating costs, means that system finances are viable, and create a positive feedback for expansion. There is no doubt that tramways are on the right track to transform urban areas into sustainable and environmentally healthy places to live and work. Passengers love trams. All that is needed is for politicians catch up with public opinion, and encourage new tram lines, that improve connectivity and attract car trips, reduce congestion and improve the urban atmosphere.

LJS Lesley
ISBN 978-1-83708-399-2
https://doi.org/10.1108/978-1-83708-398-520261013

References

Anderson N (2009, February) *Freight by Tram - Is There a Case for It?* Tramways & Urban Transit, No. 854.

Atkins HN (1980) *Highway Materials, Soils and Concretes*. Reston Publishing, VA, USA.

Audit Commission (2006, November) *Annual Audit Letter.* merseytravel.gov.uk/pdf/MerseytravelAAL.pdf (accessed 01/05/2026).

Audit Scotland, Interim Report (2011) https://audit.scot/publications/edinburgh-trams-interim-report (accessed 01/05/2026).

Bae CHC, Jun M-J and Park H (2003) The impact of Seoul's subway Line 5 on residential property values. *Transport Policy* **10(2)**.

Banister D (2002) *Transport Planning*. Spon Press, London, UK.

Bartlett D (2008, January 21) Cash waste over Merseytram after local infighting, *Liverpool Daily Post.*

Bennett RF (1971) *Increasing the attractiveness of public transport.* UITP Congress, Rome, Italy.

BBC (2006) http://news.bbc.co.uk/1/hi/england/london/5098872.stm (accessed 01/05/2026).

Bell MG (1985) A robot street tram. *Light Rail Transport Proceedings*, Liverpool Polytechnic, Liverpool, UK. ISBN 0906442133.

Black E (2007) Chapter Ten: 'The GM Conspiracy'. In *Internal Combustion: How Corporations and Governments Addicted the World to Oil and Derailed the Alternatives*. Macmillan, pp. 193–260. ISBN 9780914153115.

Bly P (1976) *The Effect of Fares on Bus Patronage LR 773*. Transport and Road Research Laboratory, Crowthorne, Berkshire, UK.

Bonsall PW and O'Flaherty CA (1997) *Observational Traffic Surveys, Transport Planning and Traffic Engineering* (O'Flaherty CA (ed)). Arnold, London, UK. ISBN: 9780340662793.

Bonsall PW (1997) *Principles of Transport Analysis and Forecasting, Transport Planning and Traffic Engineering* (O'Flaherty CA (ed)). Arnold, London, UK.

Boulder (2022) https://boulderco;orado.gov/news/2022-boulder-valley-employee-survey-travel-findings-published (accessed 01/05/2026).

Bruton MJ (1970) *Introduction to Transportation Planning*. Hutchinson, London, UK. ISBN 9781003155690.

Buchanan CD (1963) *Traffic in Towns, Report of Working Group*. HMSO, London, UK.

Buckley RJ (1975) *History of Tramways from Horse to Rapid Transit*. David and Charles, Exeter, UK. ISBN: 0715366416.

Butler EW, Chapin FS, Hemmens G, Kaiser EJ Steginan MA and Weiss SF (1969) *Moving behavior and residential choice*. Highway Research Board, Special Report 81.

Bygate J (1995) South Yorkshire Supertram – first year of operation. *Light Rail 95 Conference*, Croydon, London, UK, 14–16th November.

CBS News (2026) https://www.cbsnews.com/atlanta/news/atlanta-ranks-among-the-worst-in-the-nation-or-traffic-bottlenecks-new-report-finds/ (accessed 01/05/2026).

Chase C (1998, September 2) Jolly trolley folly. *San Diego Earth Times*. www.sdearthtimes.com/cut_to_chase/ctc_on_trans.html (accessed 01/05/2026).

Cheek C (2008, August) NTS data reinforces effects of car ownership on bus use. *Transit Magazine*.

Christaller W (1933) *Die zentralen Orte in Suddeutschland. Jena: Gustav Fischer, 1933*. (Translated (in part), by Charlisle W Baskin (1966), as Central Places in Southern Germany). Prentice Hall, NJ, USA.

Churchman GB (1995) *Railway Electrification in Australia and New Zealand*. IPL Books, London, UK.

Coffey A and Kuchwalek H (1992) *Grunes licht fur die strassenbahn*. Verlag fur Gesellschaftskritik, Vienna, Austria.

Collins P (2001) *London's Trams – A View from the Past*. Ian Allan, Shepperton, UK.

Cudahy BJ (1995) *Cash, Tokens and Transfers*. Fordham University Press, NY, USA.

Cuffe P (1989) Control and Communications. *Light Rail 89 Proceedings*, Liverpool Polytechnic, Liverpool, UK.

Daly A (1981) *Behavioural Travel Modelling, Transport & Public Policy Planning*. Mansell, London, UK.

Davis, WJ (1926) The tractive resistance of electric locomotives and cars. *General Electric Review* **29**.

Dewees DN (1976) The effect of a subway on residential property values in Toronto, Elsevier. *Journal of Urban Economics* **3(4)**.

Dix M (1981) *Contributions of Psychology to Developments in Travel Demand Modelling, Transport & Public Policy Planning*. Mansell, London, UK.

Dukkipati R (2000) *Vehicle Dynamics*. Narosa Publishing House, Delhi, India. pp. 227–228.

Dunbar CS (1967) *Buses, Trolleys & Trams*. Paul Hamlyn, London, UK.

Ellen, ER and Phillips I (1973) *Financing the Revenue and Capital Costs of Public Transport Undertakings*. 40th International Congress.

EU Urban Mobility Observatory (2021) Free passenger transport - exploring the benefits and disadvantages. https://urban-mobility-observatory.transport.ec.europa.eu/resources/case-studies/free-passenger-transport-exploring-benefits-and-disadvantages_en (accessed 01/05/2026).

Faludi AKF (2004) Impact of a planning philosophy. *Planning Theory* **3(3)**: 225–236.

Fast Company (2023) Estonia's capital made mass transit free a decade ago. Car traffic went up. https://www.fastcompany.com/90968891/estonias-capital-made-mass-transit-free-a-decade-ago-car-traffic-went-up (accessed 01/05/2026).

Faulkner P (2020) Preston tram service could be on track for 2022 launch after test runs approved. Lancashire Evening Post, 3rd December 2020 https://www.lep.co.uk/news/transport/preston-tram-service-could-be-on-track-for-2022-launch-after-test-runs-approved-3056897 (accessed 01/05/2026).

Fernandez V (2023) Transport Heritage Expo. https://www.instagram.com/vfernandez_299/p/CzOO5qtPTuK/?img_index=1 (accessed 01/05/2026).

Garbutt P (1989) *World Metro Systems*. Capital Transport, Harrow, UK.

Glaister S (1981) *Fundamentals of Transport Economics*. Wiley Blackwell, NJ, USA.

Golob TF, Canty ET and Gustafson RL (1972, March) *Transportation Research* **6(1)**.

Goodwin P, *et al.* (1983) *Subsidised Public Transport and the Demand for Travel*. The South Yorkshire Example, TRRL Report, ISBN: 0-566-00654-5.

Goodwin PB (1993) Car ownership and public transport use: Revisiting the interaction. *Transportation* **20(1)**, ISSN 0049 4488.

Guardian (2024) Fare-free public transport: good for people as well as the planet. https://www.theguardian.com/commentisfree/2024/jan/07/the-guardian-view-on-fare-free-public-transport-good-for-people-as-well-as-the-planet (accessed 01/05/2026).

Gutnecht R (1973) Alternative approaches to public transport and the implications for revenue and traffic. *UITP 40th Congress.*

Hardie Report (2023), Edinburgh Tram Inquiry, https://www.edinburghtraminquiry.org (accessed 01/05/2026).

Hass-Klau C (2002) *Future of Urban Transport.* ETP.

Hass-Klau C (2003) *Bus or Light Rail: Making the Right Choice*, 2nd edn., ETP.

Hass-Klau C (2004) *Economic Impact of Light Rail: The Results of 15 Urban Areas in France, Germany, UK and North America.* ETP.

Hay A (1986) The impact of subsidised low-fare public transport on travel behaviour. *Environment and Planning C: Government and Policy* **4(2)**: 233–246.

Hay A (1986) *The Impact of Subsidised Low Fare Public Transport on Travel Behaviour.* TRRL RR38.

Hennessey RAS (2008, July) Sparks: the electrical consultants. *BackTrack* **22(7)**:390–396.

Hensher DA (1974) *Consumer Preferences in Urban Trip-Making.* Commonwealth Bureau of |Roads, Australia.

Hensher DA (1989) Behavioural and resource values of travel time savings. *Australian Road Research* **19(3)**.

Hills PJ (1981) Does urban bus transport have a future? *Transportation* **10(4)**.

HMG Transport Act (1985) https://www.legislation.gov.uk/ukpga/1985/67/contents (accessed 01/05/2026).

Hondius H (2007) Combine rectification programme moves ahead. *Railway Gazetter International*, May 1 2007.

Horne JB and Maund TB (1982) *Liverpool Transport* **2**: 1900–1930. Transport Publishing Company, Derbyshire, UK.

Horvath A, Kerkapoly E and Megyeri J (1977) *Kulonleges vasutak, Muszaki Konyvkiado*, Budapest, Hungary.

Hovell PJ, Jones WH and Moran AJ (1975) *The Management of Urban Public Transport.* Saxon House, Franborough, UK.

Hubbell J and Colquhoun D (2006) *Light Rail Transport in Calgary, the first 25 years.* Presented at the TRB International Light Rail Conference, St. Louis, Missouri, USA, 8-12 April.

Hupkes G (1982) The law of constant travel time and trip rates. *Futures* **14**: 42.

Jaurinensis (2009) Szechenyi Istvan University. ISSN 1 789-6932.

Jones B (2007, June) *South Yorkshire's Quiet Success Story.* Tramways and Urban Transit, No. 834.

Joyce J (1964) *Modern Tramway Review.* Ian Allan, Shepperton, UK.

Kappis C and Schreiter H (2016) Handbook Track Greening (PDF). Eurail Press, Hamburg, Germany.

Keenan D (1979) *Tramways of Sydney.* Transit Press.

Knupfer S, Pokotilo V and Woetzel J (2018) *Urban Transportation Systems of 24 World Cities.* McKinsey and Co. https://www.mckinsey.com/~/media/McKinsey/Business%20Functions/Sustainability/Our%20Insights/Elements%20of%20success%20Urban%20transportation%20systems%20of%2024%20global%20cities/Urban-transportation-systems_e-versions.ashx (accessed 01/05/2026).

Laconte P (1978) *Marketing and Public Transport.* UITP Conference, Newcastle upon Tyne, UK.

Langdon MG (1971) *Cabtrack Studies, Royal Aircraft Establishment.* Ministry of Defence, London, UK.

Latour B (1971) The Cabtrack urban transport system. *Traffic Engineering & Control* **2**: 634–638.
Legris R (1971) *Measuring and Improving Productivity in Urban Passenger Surface Transport.* Paper No. 5, UITP Congress, Rome.
Lesley L (1972) Synthesising networks. *Traffic Engineering and Control* 1972(11).
Lesley L (1973) Determining the optimum number of interchange stations for a city. *PTRC Winter Conference*. Sussex University.
Lesley L (1974, August) *The Effect of Advertising Bus Services in Two Housing Estates in the Teesside Conurbation.* Action Report No. 5, University of Newcastle upon Tyne.
Lesley L (1974, August) *The image of public transport in two housing estates in the Teesside conurbation.* Action Report No.4, University of Newcastle upon Tyne.
Lesley L (1974, March 01) *Improving the Image, Motor Transport, Bus & Coach Supplement.*
Lesley L (1976) Optimum bus stop spacing. Traffic Engineering and Control 1976/10 part 1 and 1976/11 part 2.
Lesley L (1983) Runcorn – a rapid transit new town. *Built Environment* **9(3/4)**.
Lesley L (1986) Public transport in Budapest – some observations about future developments. *Periodica Polytechnica* **14(1)**. Budapest University of Technology and Economics.
Lesley L (1987, November) *Budget Light Rail – A Product of the Discount Rate, Conference Proceedings, Electrifying Urban Public Transport.* Liverpool Polytechnic, Liverpool, UK.
Lesley L (1991) *Light Rail Handbook.* Transport Science Ltd.
Lesley L (1996) *Public & Private Funding in New Urban Public Transport, Transport in Transition.* Public Finance Foundation.
Lesley L (2002, July 3) Testing the next generation of tramway tracks. *Railway Engineering 2002 Conference*, London, UK.
Lesley L (2005, February) *Affordable Mass Transit?* Mass Transit, Washington DC.
Lesley L (2006, October) *New OHL system*. The Rail Engineer, No. 24.
Lesley L (2009) *Fatigue in Railway and Tramway Track, in Fatigue in Railway Infrastructure.* Woodhead Publishing.
Lesley L (2009) Generalised costs and the value of time as a method of patronage forecasting. *Acta Technica Jaurinensis* **2(1)**.
Lesley L (2019, June) Behaviour and limitations of 4 wheel trucks. *Railway Engineering Conference*, Edinburgh.
Lesley L (2023) Embedded Tracks. Railway Engineering Conference, Edinburgh 14th June 2023.
Lesley L (2025a) *Embedded rail tracks*, Elsevier, Amsterdam, the Netherlands.
Lesley L (2025b) Monomotor Bogies. *Tramway Review* **38(284)**.
Litterbins (2020) https://www.litterbins.co.uk/blogs/news/the-worlds-greenest-cities-freiburg-germany?srsltid=AfmBOor8wnIEFGCU8lN0NmNhcme-eBb-HcWxEopMIMI-dDe_LzZwv64t (accessed 01/05/2026).
London (2017) https://www.roads.org.uk/ringways/ringway1 (accessed 01/05/2026).
Lowson MV and Cabtrack (1999) *Proceedings Institution of Civil Engineers Transport,* **135(3)**: 139–151.
Marohn CL (2020) *Strong Towns*. John Wiley and Sons, NY, USA.
Matsoukis EC (1980) Speed/flow relationships in the central Glasgow area. *Transportation Planning and Technology* **6(1)**: 21–26, 1029-0354.
McLuhan M (1967) *The Medium Is the Message*. University of Toronto Press, Canada.
McMillan RK and Assael H (1968/9) *National Survey of Transportation Attitudes and Behaviour.* Highway Research Board, Reports No.49 and 82, Washington.
Mogridge MJH (1990) *Travel in Towns. Jam yesterday, jam today and jam tomorrow.* Palgrave MacMillan Press, London, UK.
Morgan B (1973) *Railways: Civil Engineering*. Arrow, London, UK.

National Audit Office (NAO) (2004) *Improving Public Transport in England Through Light Rail*. NAO, England.

O'Flaherty CA (1997) *Transport Planning and Traffic Engineering*. Arnold, London, UK.

Paine FT, Nash AN, Hille SJ and Brunner GA (1969) *Journal of Applied Psychology* **53(6)**.

Pas EI and Harvey AS (1997) Time use research & travel analysis modelling. In *Understanding Travel Behaviour in an Era of Change*. Pergamon, Oxford, UK.

Paulley N, Mackett R, Preston J, Wardman M, Titheridge H and White P (2004) Factors affecting the demand for public transport. *European Transport Conference*.

Pharoah T and Apel D (1995) *Transport Concepts in European Cities*. Ashgate Publishing, Surrey, UK.

Pitrel J (1987) Nantes modern tramway–first results and forecasts. *Proceedings Electrifying Urban Public Transport, Conference in Liverpool.* ISBN 0 906442 15 X, 17/18th Nov.

Powell MF (1997) *Light Rail & Guided Transit*. Natula Publications, Dorset, UK.

Profillidis VA (2000) *Railway Engineering*. Ashgate Publishing, Surrey, UK.

Salkield T (1953) *Road Making and Road Using*. Pitman, London, UK.

Sharp DR (1970) *Concrete in Highway Engineering*. Pergamon Press, Oxford, UK.

Skamris M and Flyvbjerg B (1997) Inaccuracy of traffic forecasts and cost estimates on large transport projects. *Transportation Policy* **4(3)**.

Slinn M, Matthews P and Guest P (1998) *Traffic Engineering Design*. Arnold, ISBN 0 340 67647 7.

Snowdon J (2007, August) *Croydon: Track Renewal Breaks the Mould*, Tramway and Urban Transit, No. 836.

Spear BD (1976). Attitudinal modeling: its role in travel-demand forecasting. In *Behavioral Travel-Demand Models*.

Stanley JK (1977) Behavioral modeling: and evaluative perspective. *3rd International Conference on Behavioral Modeling*, South Australia.

Starkie DNM (1976) *Transportation Planning, Policy and Analysis*. Pergamon Press, Oxford, UK.

Steimel A (2007) *Motive power and energy supply: basics and practical experience*. Oldenbourg Industrieverlag, Munich, Germany.

Streeter T (2009, May) *Dresden: Innovation Is Born from Experience*. Tramways & Urban Transit, No. 857.

SYMCA (2024) https://www.southyorkshire-ca.gov.uk/news/article/fare-freeze (accessed 01/05/2026).

Taylor MAP (2024) Distribution of walking access to public transport in Melbourne, Australia-Evidence on acceptable and tolerable walking distances. *International Journal of Sustainable Transportation* **18(7)**.

Tennyson EL (1989) Impact on Transit Patronage of Cessation or Inauguration of Rail Service. TRB Report 1221.

Tolley R and Turton B (1995) *Transport Systems, Policy and Planning*. Longman, London, UK.

Townsend EC (1969) *Investment and Uncertainty*. Oliver and Boyd, Edingburgh, UK.

Transportation Research Board (1989) *Transportation Research Record 1221*. Transportation Research Board, Washington DC, USA.

Transportation Research Board (1992) *Transportation Research Record No. 1361*, Light Rail Transit: Planning, Design, and Operating Experience (Papers presented at the Sixth National Conference on Light Rail Transit, held in Calgary, Alberta, Canada, May 24-27, 1992).

Transportation Research Board (2009, March) *Track Design Handbook for Light Rail Transit, TCRP D-14*, Washington, USA.

UITP (Union Internationale des Transports Publics) (1991) *Ten Case Studies, Light Rail Commission Paper*, 49th Congress, Stockholm, Sweden.

UITP (2019) The global tram and light rail landscape 2019–2021, Report https://www.uitp.org/publications/the-global-tram-and-light-rail-landscape-2019-21 (accessed 01/05/2026).
UITP 49th Congress (1991) *Ten Case Studies, Light Rail Commission Paper*. Stockholm, Sweden.
UITP Light Rail Report. https://www.uitp.org/topics/light-rail/ (accessed 01/05.2026).
UITP Tram/LRT London city transit data https://citytransit.uitp.org/london/public-transport-ridership/tram-lrt (accessed 01/05/2026).
UK DfT (Department for Transport) (2025) https://www.gov.uk/government/statistics/concessionary-travel-statistics-year-ending-march-2025/concessionary-travel-statistics-year-ending-march-2025 (accessed 01/05/2026).
UNCTAD (United Nations Trade and Development) (2022) *Handbook of Statistics 2022*. United Nations Publications, Geneva, Switzerland.
Van der Bijl, R, Nam Oort, N and Nukman, B (2018) *Light Rail Transit Systems*. Elsevier, Amsterdam, the Netherlands.
US DoT (Department of Transportation) (1976) *Effective Citizen Participation in Transportatioin Planning*, **2** vols. US Government Printing Office, Washington DC, USA.
Vickerman RW *Infrastructure and Regional Development*. Pion Press, London, UK.
Voorhees AM and Associates (1974, July) *Public transport in Bedfordshire: Facts, Views, Issues, Options*. Alan Voorhees and Associates.
VVR (2026) https://www.vrr.de (accessed 01/05/26)
Wachs M (1985) The politicization of transit subsidy policy in America. In *Transportation and Mobility in an Era of Transition*. Elsevier, Amsterdam, the Netherlands.
Webster FV (1968) *A theoretical estimate of London car commuters transferring to bus travel*. RRL LR 165.
Webster FV (1977) *Urban passenger transport, some trends and prospects*. TRRL LR 771.
Wendell W (2008) Bus rapid transit, Breakthrough Technologies Institute. https://www.scirp.org/reference/referencespapers?referenceid=2052128 (accessed 01/05/26).
White P (2005) *Public Transport, Planning, Management & Operation*. UCL Press, London, UK.
WHO (2017) Guidelines for noise. https://cpcb.nic.in/who-guidelines-for-noise-quality/ (accessed 01/05/2026).
WHO (n.d.) Air quality guidelines. https://www.eea.europa.eu/publications/status-of-air-quality-in-Europe-2022/europes-air-quality-status-2022/world-health-organization-who-air (accessed 01/05/2026).
Willmore PL (1967) Transportation Hiercharchies. *Architectural Review* **142(189)**.
Wondertram (2025) https://wondertram.com/wonder-track-technology/ (accessed 01/05/2026).
Yearsley I (1993) *Previously unexplored aspects of London's tramway finances*. Tramway, London, UK.

WEBSITES

www.apta.com/sites/transus/lightrail.htm
www.lightrail.com
www.lightrail.net
www.lightrailnow.org
www.LRTA.org
www.transport2000.ca
www.vtpi.org
www.wikipedia.org/wiki/Light_rail

LJS Lesley
ISBN 978-1-83708-399-2
https://doi.org/10.1108/978-1-83708-398-520261011

Index

Printed and bound by CPI Group (UK) Ltd, Croydon, CR0 4YY

16/08/2026

14942380-0001